Iosias Jody (Ed.)

Lake Henshaw

Iosias Jody (Ed.)

Lake Henshaw

Reservoir, Bank (geography), Llwyn-on Reservoir, North County, San Diego

Cred Press

Imprint

All parts of this book are extracted from Wikipedia, the free encyclopedia (www.wikipedia.org).

You can get detailed informations about the authors of this collection of articles at the end of this book. The editors (Ed.) of this book are no authors. They have not modified or extended the original texts.

Pictures published in this book can be under different licences than the GNU Free Documentation License. You can get detailed informations about the authors and licences of pictures at the end of this book.

The content of this book was generated collaboratively by volunteers. Please be advised that nothing found here has necessarily been reviewed by people with the expertise required to provide you with complete, accurate or reliable information. Some information in this book maybe misleading or wrong. The Publisher does not guarantee the validity of the information found here. If you need specific advice (f.e. in fields of medical, legal, financial, or risk management questions) please contact a professional who is licensed or knowledgeable in that area.

Cover image: www.ingimage.com
Concerning the licence of the cover image please contact ingimage.

Publisher:
Cred Press is a trademark of
International Book Market Service Ltd., 17 Rue Meldrum, Beau Bassin, 1713-01 Mauritius
Email: info@bookmarketservice.com
Website: www.bookmarketservice.com

Published in 2011

Printed in: U.S.A., U.K., Germany. This book was not produced in Mauritius.

ISBN: 978-613-7-39671-1

Contents

Articles

References

Reservoir

A **reservoir** (pronounced either REZ-ur-vore or REZ-er-vwar, etymology: from French *réservoir* a "storehouse [1]) or an **artificial lake** is used to store water. Reservoirs may be created in river valleys by the construction of a dam or may be built by excavation in the ground or by conventional construction techniques such as brickwork or cast concrete.

The Jhonghua Dam on the Dahan River in Taoyuan County, Taiwan.

The term reservoir may also be used to describe underground reservoirs such as an oil or water well.

Types

Valley dammed reservoir

A dam constructed in a valley relies on the natural topography to provide most of the basin of the reservoir. Dams are typically located at a narrow part of a valley downstream of a natural basin. The valley sides act as natural walls with the dam located at the narrowest practical point to provide strength and the lowest practical cost of construction. In many reservoir construction projects people have to be moved and re-housed, historical artifacts moved or rare environments relocated. Examples include the temples of Abu Simbel[2] (which were moved before the construction of the Aswan Dam to create Lake Nasser from the Nile in Egypt) and the re-location of the village of Capel Celyn during the construction of Llyn Celyn.[3]

Lake Vyrnwy Reservoir. The dam spans the Vyrnwy Valley and was the first large stone dam built in the United Kingdom.

Construction of a reservoir in a valley will usually necessitate the diversion of the river during part of the build often through a temporary tunnel or by-pass channel.[4]

In hilly regions reservoirs are often constructed by enlarging existing lakes. Sometimes in such reservoirs the new top water level exceeds the watershed height on one or more of the feeder streams such as at Llyn Clywedog in Mid Wales.[5] In such cases additional side dams are required to contain the reservoir.

Stocks Reservoir in Lancashire, England.

Where the topography is poorly suited to a single large reservoir, a number of smaller reservoirs may be constructed in a chain such as in the River Taff valley where the three reservoirs Llwyn-on Reservoir, Cantref Reservoir and Beacons Reservoir form a chain up the valley.[6]

Bank-side reservoir

Where water is taken from a river of variable quality or quantity, bank-side reservoirs may be constructed to store the water pumped or siphoned from the river. Such reservoirs are usually built partly by excavation and partly by the construction of a complete encircling bund or embankment which may exceed 6 km in circumference.[7] Both the floor of the reservoir and the bund must have an impermeable lining or core, often made of puddled clay. The water stored in such reservoirs may have a residence time of several months during which time normal biological processes are able to substantially reduce many contaminants and almost eliminate any turbidity. The use of bank-side reservoirs also allows a water abstraction to be closed down for extended period at times when the river is unacceptably polluted or when flow conditions are very low due to drought. The London water supply system is one example of the use of bank-side storage for all the water taken from the River Thames and River Lee with many large reservoirs such as Queen Mary Reservoir visible along the approach to London Heathrow Airport.[7]

Service reservoir

Service reservoirs[8] store fully treated potable water close to the point of distribution. Many service reservoirs are constructed as water towers, often as elevated structures on concrete pillars where the landscape is relatively flat. Other service reservoirs are entirely underground, especially in more hilly or mountainous country. In the United Kingdom, Thames Water has many underground reservoirs built in the 1800s by the Victorians, most of which are lined with brick. A good example is the Honor Oak Reservoir, constructed between 1901 and 1909. When it was completed it was the largest brick built underground reservoir in the world[9] and is still one of the largest in Europe[10] . The reservoir now forms part of the Southern extension of the Thames Water Ring Main. The top of the reservoir has been grassed over and is now the Aquarias Golf Club[11] .

Service reservoirs perform several functions including ensuring sufficient head of water in the water distribution system and providing hydraulic capacitance in the system to even out peak demand from consumers enabling the treatment plant to run at optimum efficiency. Large service reservoirs can also be managed to so that energy costs in pumping are reduced by concentrating refilling activity at times of day when power costs are low.

History

Five thousand years ago, the craters of extinct volcanoes in Arabia were used as reservoirs by farmers for their irrigation water.[12]

Dry climate and water scarcity in India led to early development of water management techniques, including the building of a reservoir at Girnar in 3000 BC.[13] Artificial lakes dating to the 5th century BC have been found in ancient Greece.[14] An artificial lake in present-day Madhya Pradesh province of India, constructed in the 11th century, covered 650 square metres (7000 sq ft).[13]

In Sri Lanka large reservoirs have been created by ancient Sinhalese kings in order to save the water for irrigation. The famous Sri Lankan king Parākramabāhu I of Sri Lanka stated " do not let a drop of water seep into the ocean without benefiting mankind ". He created the reservoir named Parakrama Samudra(sea of King Parakrama),[15] which has astonished archaeologists.

Uses

Direct water supply

Gibson Reservoir, Montana

Many dammed river reservoirs and most bank-side reservoirs are used to provide the raw water feed to a water treatment plant which delivers drinking water through water mains. The reservoir does not simply hold water until it is needed; it can also be the first part of the water treatment process. The time the water is held for before it is released is known as the *retention time*. This is a design feature that allows particles and silts to settle out, as well as time for natural biological treatment using algae, bacteria and zooplankton that naturally live within the water. However natural limnological processes in temperate climate lakes produces temperature stratification in the water body which tends to partition some elements such as manganese and phosphorus into deep, cold anoxic water during the summer months. In the autumn and winter the lake becomes fully mixed again. During drought conditions, it is sometimes necessary to draw down the cold bottom water and the elevated levels of manganese in particular can cause problems in water treatment plants.[16]

Hydroelectricity

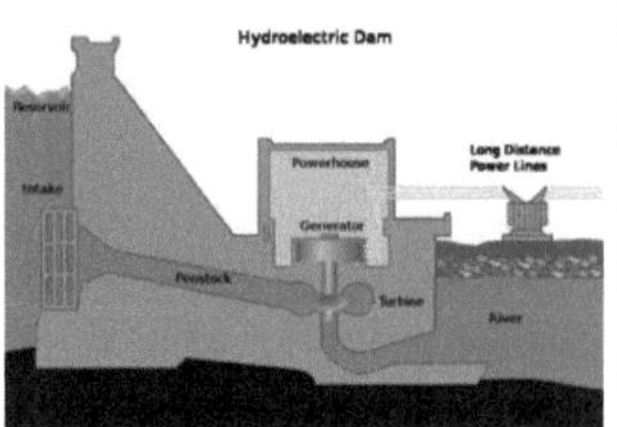

Hydroelectric dam in cross section.

A reservoir generating hydroelectricity includes turbines connected to the retained water body by large-diameter pipes. These generating sets may be at the base of the dam or some distance away. Some reservoirs generating hydroelectricity use pumped re-charge in which a high-level reservoir is filled with water using high-performance electric pumps at times when electricity demand is low and then uses this stored water to generate electricity by releasing the stored water into a low-level reservoir when electricity demand is high. Such systems are called pump-storage schemes.[17]

Controlling watercourses

Reservoirs can be used in a number of ways to control how water flows through downstream waterways.

Downstream water supply – water may be released from an upland reservoir so that it can be abstracted for drinking water lower down the system, sometimes hundred of miles further down downstream

Irrigation – water in an irrigation reservoir may be released into networks of canals for use in farmlands or secondary water systems. Irrigation may also be supported by reservoirs which maintain river flows allowing water to be abstracted for irrigation lower down the river.[18]

Flood control – also known as an "*attenuation*" or "*balancing*" reservoir, flood control reservoirs collect water at times of very high rainfall, then release it slowly over the course of the following weeks or months. Some of these reservoirs are constructed across the river line with the onward flow controlled by an orifice plate. When river flow exceeds the capacity of the orifice plate water builds behind the dam but as soon as the flow rate reduces the water behind the dam slowly releases until the reservoir is empty again. In some cases such reservoirs only function a few times in a decade and the land behind the reservoir may be developed as community or recreational land. A new generation of balancing dams are being developed to combat the climatic consequences of climate change. They are called "Flood Detention Reservoirs". Because these reservoirs will remain dry for long periods, there may be a risk of the clay core drying out reducing its structural stability. Recent developments include the use of composite core fill made from recycled materials as an alternative to clay.

Canals – Where a natural watercourse's water is not available to be diverted into a canal, a reservoir may be built to guarantee the water level in the canal; for example, where a canal climbs to cross a range of hills through locks.[19]

Recreation – water may be released from a reservoir to artificially create or supplement white-water conditions for kayaking and other white-water sports.[20] On salmonid rivers special releases (in Britain called *freshets*) are made to encourage natural migration behaviours in fish and to provide a variety of fishing conditions for anglers.

Recreational-only Kupferbach reservoir near Aachen/Germany.

Flow balancing

Reservoirs can be used to balance the flow in highly managed systems, taking in water during high flows and releasing it again during low flows. In order for this to work without pumping requires careful control of water levels using adjustable sluices. Accurate weather forecasts are essential so that dam operators can plan drawdowns prior to a high rainfall or snowmelt event. Dam operators blamed a faulty weather forecast on the 2010–2011 Queensland floods. Examples of highly managed Reservoirs are Burrendong Dam in Australia and Llyn Tegid in North Wales. Llyn Tegid is a natural lake whose level was raised by a low dam and into which the River Dee flows or discharges depending upon flow conditions at the time as part of the River Dee regulation system. This mode of operation is a form of hydraulic capacitance in the river system.

Recreation

The water bodies provided by many reservoirs often allow some recreational uses such as fishing, boating, and other activities. Special rules may apply for the safety of the public and to protect the quality of the water and the ecology of the surrounding area. Many reservoirs now support and encourage less informal and less structured recreation such as natural history, bird watching, landscape painting, walking and hiking and often provide information boards and interpretation material to encourage responsible use.

Operation

Water falling as rain upstream of the reservoir together with any groundwater emerging as springs is stored in the reservoir. Any excess water can be spilled via a specifically designed spillway. Stored water may be piped by gravity for use as drinking water, to generate hydro-electricity or to maintain river flows to support downstream uses. Occasionally reservoirs can be managed to retain high rain-fall events to prevent or reduce downstream flooding. Some reservoirs support several uses and the operating rules may be complex.

Most modern reservoirs have a specially designed draw-off tower that can discharge water from the reservoir at different levels both to access water as the reservoir draws down but also to allow water of a specific quality to be discharged into the downstream river as compensation water.

The operators of many upland or in-river reservoirs have obligations to release water into the downstream river to maintain river quality, support fisheries, maintain downstream industrial uses. maintain recreational use or for a range of other requirements. Such releases are known as *compensation water.*

Spillway of Llyn Brianne dam in Wales.

Terminology

The terminology for reservoirs varies from country to country. In most of the world reservoir areas are expressed in km^2 whilst in the USA acres are commonly used. For volume either m^3 or km^3 are widely used with acre feet used in the USA.

The capacity, volume or storage of a reservoir is usually divided into distinguishable areas. *Dead* or *inactive* storage refers to water in a reservoir that cannot be drained by gravity through a dam's outlet works, spillway or power plant intake and can only be pumped out. Dead storage allows sediments to settle which improves water quality and also creates hydraulic head along with an area for fish during low levels. *Active* or *live* storage is the portion of the reservoir that can be utilized for flood control, power production, navigation and downstream releases. In addition, a reservoir's *flood control capacity* is the amount of water it can regulate during flooding. The *surcharge capacity* is the capacity of the reservoir above the spillway crest that cannot be regulated.[21]

In the United States the water below the normal maximum level of a reservoir is called the *conservation pool.*[22]

In the UK *top water level* describes the reservoir full state whist *fully drawn down* describes the minimum retained volume.

Modelling reservoir management

There is a wide variety of software for modelling reservoirs, from the specialist Dam Safety Program Management Tools (DSPMT) to the relatively simple WAFLEX, to integrated models like the Water Evaluation And Planning system (WEAP) that place reservoir operations in the context of system-wide demands and supplies.

Safety

In many countries large reservoirs are closely regulated to try to prevent or minimise failures of containment.[23] [24]

Whilst much of the effort is directed at the dam and its associated structures as the weakest part of the overall structure, the aim of such controls is to prevent an uncontrolled release of water from the reservoir. Reservoir failures can generate huge increases in flow down a river valley with the potential to wash away towns and villages and cause considerable loss of life such as the devastation following the failure of containment at Llyn Eigiau which killed 17 people.[25] (see also List of dam failures)

A notable case of reservoirs being used as an instrument of War involved the British Royal Air Force Dambusters raid on Germany in World War II (codenamed "Operation Chastise" [26]), in which three German reservoir dams were selected to be breached in order to impact on German infrastructure and manufacturing and power capabilities deriving from the Ruhr and Eder rivers. The economic and social impact was derived from the enormous volumes of previously stored water that swept down the valleys wreaking destruction. This raid later became the basis for several films.

Environmental impact

Whole life environmental impact

All reservoirs will have a monetary cost/benefit assessment made before construction to see if the project is worth proceeding with.[27] However, such analysis can often omit the environmental impacts of dams and the reservoirs that they contain. Some impacts such as the greenhouse gas production associated with concrete manufacture are relatively easy to estimate. Other impact on the natural environment and social and cultural effects can be more difficult to assess and to weigh in the balance but identification and quantification of these issues are now commonly required in major construction projects in the developed world [28]

Climate change

Depending upon the circumstances, a reservoir built for hydro-electricity generation can either reduce or increase the net production of greenhouse gases. An **increase** can occur if plant material in the flooded areas decays in an anaerobic environment releasing (methane and carbon dioxide). This apparently counterintuitive position arises because much carbon is released as methane which is approximately 8 time more potent as a greenhouse gas than carbon dioxide [29]

A study for the National Institute for Research in the Amazon found that Hydroelectric reservoirs release a large pulse of carbon dioxide from above-water decay of trees left standing in the reservoirs, especially during the first decade after closing.[30] This elevates the global warming impact of the dams to levels much higher than would occur by generating the same power from fossil fuels.[30] According to the World Commission on Dams report (Dams And Development), when the reservoir is relatively large and no prior clearing of forest in the flooded area was undertaken, greenhouse gas emissions from the reservoir could be higher than those of a conventional oil-fired thermal generation plant.[31] For instance, In 1990, the impoundment behind the Balbina Dam in Brazil (inaugurated in 1987) had over 20 times the impact on global warming than would generating the same power from fossil fuels, due to the large area flooded per unit of electricity generated.[30]

A **decrease** can occur if the dam is used in place of traditional power generation, since electricity produced from hydroelectric generation does not give rise to any flue gas emissions from fossil fuel combustion (including sulfur dioxide, nitric oxide and carbon monoxide from coal). The Tucurui dam in Brazil (closed in 1984) had only 0.4 times the impact on global warming than would generating the same power from fossil fuels.[30]

Biology

Dams can produce a block for migrating fish, trapping them in one area, producing food and a habitat for various water-birds. They can also flood various ecosystems on land and may cause extinctions.

Human Impact

Dams can severely reduce the amount of water reaching countries downstream of them, causing water stress between the countries, e.g. the Sudan and Egypt, which damages farming businesses in the downstream countries, and reduces drinking water.

Farms and villages, e.g. Ashopton can be flooded by the creation of reservoirs, ruining many livelihoods. For this very reason, worldwide 80 million people (figure is as of 2009) have had to be forcibly relocated due to dam construction.

Limnology

The limnology of reservoirs has many similarities to that of lakes of equivalent size. There are however significant differences.[32] Many reservoirs experience considerable variations in level producing significant areas that are intermittently underwater or dried out. This greatly limits the productivity or the water margins and limits the number of species able to survive in these conditions.

Upland reservoirs tend to have a much shorter residence time than natural lakes and this can lead to more rapid cycling of nutrients through the water body so that they are more quickly lost to the system. This may be seen as a mismatch between water chemistry and water biology with a tendency for the biological component to be more oligotrophic than the chemistry would suggest.

Conversely, lowland reservoirs drawing water from nutrient rich rivers, may show exaggerated eutrophic characteristics because the residence time in the reservoir is much greater than in the river and the biological systems have a much greater opportunity to utilise the available nutrients.

Deep reservoirs with multiple level draw off towers can discharge deep cold water into the downstream river greatly reducing the size of any hypolimnion. This in turn can reduce the concentrations of phosphorus released during any annual mixing event and may therefore reduce productivity.

The Dams in front of reservoirs act as knickpoints-the energy of the water falling from them reduces and deposition is a result below the Dams.

Seismicity

The filling (impounding) of reservoirs has often been attributed to reservoir-triggered seismicity (RTS) as seismic events have occurred near large dams or within their reservoirs in the past. These events may have been triggered by the filling or operation of the reservoir and are on a small scale when compared to the amount of reservoirs worldwide. Of over 100 recorded events, early examples include the 60 m (197 ft) tall Marathon Dam in Greece (1929), the 221 m (725 ft) tall Hoover Dam in the U.S. (1935). Most events involve large dams and small amounts of seismicity. The only four recorded events above a 6.0-magnitude (M_w) are the 103 m (338 ft) tall Koyna Dam in India which registered a M_w of 6.3 along with the 120 m (394 ft) Kremasta Dam in Greece which registered a 6.3-M_w as well. Following those two, the next largest were the 122 m (400 ft) high Kariba Dam in Zambia at 6.25-M_w and the 105 m (344 ft) Xinfengjiang Dam in China at 6.1-M_w. Disputes occur over when RTS has occurred due to a lack of hydrogeological knowledge at the time of the event. It is accepted though that the infiltration of water into pores and the weight of the reservoir do contribute to RTS patterns. For RTS to occur, there must be a seismic structure near the dam or its reservoir and the seismic structure must be close to failure. Additionally, water must be able to infiltrate the deep rock stratum as the weight of a 100 m (328 ft) deep reservoir will have little impact when compared the deadweight of rock on a crustal stress field which may be located at a depth of 10 km (6 mi) or more.[33]

Micro climate

Reservoirs may change the local micro-climate increasing humidity and reducing extremes of temperature. Such effects are claimed by some South Australian winerys as increasing the quality of the wine production.

List of reservoirs

List of reservoirs by area

The following are the world's ten largest reservoirs by surface area:

1. Lake Volta (8482 km^2 or 3275 sq mi; Ghana) [34]
2. Smallwood Reservoir (6527 km^2 or 2520 sq mi; Canada)[35]
3. Kuybyshev Reservoir (6450 km^2 or 2490 sq mi; Russia)[36]
4. Lake Kariba (5580 km^2 or 2150 sq mi; Zimbabwe, Zambia)[37]
5. Bukhtarma Reservoir (5490 km^2 or 2120 sq mi; Kazakhstan)
6. Bratsk Reservoir (5426 km^2 or 2095 sq mi; Russia)[38]
7. Lake Nasser (5248 km^2 or 2026 sq mi; Egypt, Sudan) [39]
8. Rybinsk Reservoir (4580 km^2 or 1770 sq mi; Russia)
9. Caniapiscau Reservoir (4318 km^2 or 1667 sq mi; Canada)[40]
10. Lake Guri (4250 km^2 or 1640 sq mi; Venezuela)

Lake Volta from space (April 1993).

List of reservoirs by volume

1. Lake Kariba (180 km^3 or 43 cu mi; Zimbabwe, Zambia)
2. Bratsk Reservoir (169 km^3 or 41 cu mi; Russia)
3. Lake Nasser (157 km^3 or 38 cu mi; Egypt, Sudan)
4. Lake Volta (148 km^3 or 36 cu mi; Ghana)
5. Manicouagan Reservoir (142 km^3 or 34 cu mi; Canada)[41]
6. Lake Guri (135 km^3 or 32 cu mi; Venezuela)
7. Williston Lake (74 km^3 or 18 cu mi; Canada)[42]
8. Krasnoyarsk Reservoir (73 km^3 or 18 cu mi; Russia)
9. Zeya Reservoir (68 km^3 or 16 cu mi; Russia)

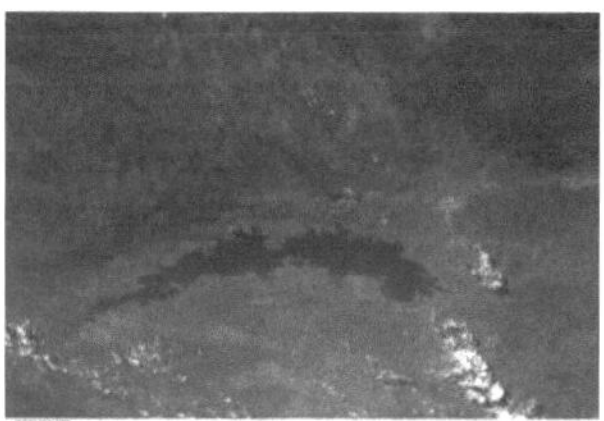

Lake Kariba from space.

See also

- Ab Anbar
- Colourful lakelets (in Poland)
- Drainage basin
- Drought
- Hydroelectricity
- Dam failure
- Mill pond
- Multipurpose reservoir
- Spillway
- Coastal sediment supply

References

[1] Online Etymology Dictionary – Reservoir (http://www.etymonline.com/index.php?term=reservoir)

[2] Nubian Monuments from Abu Simbel to Philae – UNESCO World Heritage Centre (http://whc.unesco.org/en/list/88)

[3] *Capel Celyn, Ten Years of Destruction: 1955–1965*, Thomas E., Cyhoeddiadau Barddas & Gwynedd Council, 2007, ISBN 978 1 900437 92 9

[4] Construction of Hoover Dam: a historic account prepared in cooperation with the Department of the Interior. KC Publications. 1976. ISBN 0-916122-51-4.

[5] Llyn Clywedog – Llanidloes mid-Wales (http://www.llanidloes.com/clywedog/index.html)

[6] Reservoirs of Fforest Fawr Geopark (http://www.geoparcyfforestfawr.org.uk/understanding/archaeology-industrial-heritage/reservoir-of-fforest-fawr-geopark)

[7] Queen Mary and King George V emergency draw down schemes (http://www.icevirtuallibrary.com/content/article/10.1680/dare.2009.19.2.79;jsessionid=3295v7olnvqlv.z-telford-01)

[8] Open University – Service Reservoirs (http://openlearn.open.ac.uk/mod/resource/view.php?id=185924)

[9] "Honor Oak Reservoir" (http://www.lewisham.gov.uk/SiteCollectionDocuments/ForestHillAndHonorOakSecretsLeaflet.pdf). London Borough of Lewisham. . Retrieved 2011-09-01.

[10] "Honor Oak Reservoir" (http://www.projectmanagement.mottmac.com/projects/?mode=type&id=130093). Mott MacDonald. . Retrieved 2011-09-01.

[11] Golf Club website (http://www.aquariusgolfclub.co.uk/pages.php/index.htmlAquarius)

[12] Smith, S. et al. (2006) *Water: the vital resource*, 2nd edition, Milton Keynes, The Open University

[13] edited by John C. Rodda, Lucio Ubertini. (2004). Rodda, John; Ubertini, Lucio. eds. *The Basis of Civilization – Water Science?*. International Association of Hydrological Science. ISBN 1-901502-57-0. OCLC 224463869

[14] Wilson & Wilson (2005). *Encyclopedia of Ancient Greece*. Routledge. ISBN 0415973341. pp. 8

[15] – International Lake Environment Committee – Parakrama Samudra (http://www.ilec.or.jp/database/asi/asi-45.html)

[16] Water problems – Manganese (http://www.freedrinkingwater.com/water_quality/chemical/water-problems-manganese.htm)

[17] How pump storage works (http://www.fhc.co.uk/pumped_storage.htm)

[18] Thinking about an irrigation reservoir? (http://www.ukia.org/eabooklets/EA Reservoir booklet_final.pdf)

[19] Huddersfield narrow canal reservoirs (http://www.huddersfield1.co.uk/huddersfield/narrowcanal/huddscanalres.htm)

[20] Water Release information for The River Tryweryn at the National Whitewater centre (http://www.ukrafting.co.uk/waterinfo.htm)

[21] Vojtěch Broža, Ladislav Votruba, (1989). *Water management in reservoirs* (http://books.google.com/?id=j8dIlPJITH0C&pg=PA187&lpg=PA187&dq=active+storage+reservoir#v=onepage&q=active storage reservoir&f=false). Elsevier Publishing Company. p. 187. ISBN 0-444-98933. .

[22] Lower Colorado River Authority – Water Glossary (http://www.lcra.org/water/conditions/glossary.html)

[23] North Carolina Dam safety law (http://www.dlr.enr.state.nc.us/pages/damsafetylaw1967.html)

[24] Reservoirs Act 1975 The Reservoirs Act 1975 (UK) (http://www.opsi.gov.uk/RevisedStatutes/Acts/ukpga/1975/cukpga_19750023_en_1)

[25] Snowdonia – Llyn Eigau (http://www.snowdoniaguide.com/llyn_eigiau.html)

[26] Commonwealth War Graves Commission – Operation Chastise (http://www.cwgc.org/admin/files/The Dams Raid.pdf)

[27] CIWEM – Reservoirs:Global Issues (http://www.ciwem.org/policy/policies/reservoirs.asp)

[28] Proposed reservoir – Environmental Impact Assessment (EIA) Scoping Report (http://www.whitehorsedc.gov.uk/news_views/topical_issues/xx_detailpage-4402.asp)

[29] Houghton, John (4 May 2005). "Global warming" (http://stacks.iop.org/RoPP/68/1343). *Reports on Progress in Physics* (Institute of Physics) **68** (6): 1362. doi:10.1088/0034-4885/68/6/R02. .

[30] Fearnside, P.M. 1995. hydroelectric dams in the Brazilian Amazon as sources of 'greenhouse' gases. Environmental Conservation 22(1): 7–19.]

[31] Hydroelectric power's dirty secret revealed – earth – 24 February 2005 – New Scientist (http://www.newscientist.com/article.ns?id=dn7046)

[32] Ecology of Reservoirs and Lakes (http://www.forestencyclopedia.net/p/p1483)

[33] "The relationship between large reservoirs and seismicity 08 February 2010" (http://www.waterpowermagazine.com/story.asp?storyCode=2055399). International Water Power & Dam Construction. 20 February 2010. . Retrieved 12 March 2011.

[34] International Lake Environment Committee – Volta Lake (http://www.ilec.or.jp/database/afr/afr-16.html)

[35] The Canadian Encyclopaedia – Smallwood Reservoir (http://www.thecanadianencyclopedia.com/index.cfm?PgNm=TCE&Params=A1ARTA0007465)

[36] International Lake Environment Committee – Reservoir Kuybyshev (http://www.ilec.or.jp/database/eur/deur54.html)

[37] International Lake Environment Committee – Lake Kariba (http://www.ilec.or.jp/database/afr/dafr04.html)

[38] International Lake Environment Committee – Bratskoye Reservoir (http://www.ilec.or.jp/database/asi/dasi61.html)

[39] International Lake Environment Committee – Aswam high dam reservoir (http://www.ilec.or.jp/database/afr/dafr19.html)

[40] International Lake Environment Committee – Caniapiscau Reservoir (http://www.ilec.or.jp/database/nam/dnam35.html)

[41] International Lake Environment Committee – Manicouagan Reservoir (http://www.ilec.or.jp/database/nam/dnam26.html)

[42] International Lake Environment Committee – Williston Lake (http://www.ilec.or.jp/database/nam/dnam29.html)

External links

- Department of Water Resources. "Reservoir Information" (http://cdec.water.ca.gov/misc/resinfo.html). *California Data Exchange Center*. State of California.

ltg:Iudinture

Bank (geography)

A ***geographic*** **bank** has four definitions and applications:

1. **Limnology**: The shoreline of a pond, swamp, estuary, reservoir, or lake. The grade (slope) can vary from vertical to a shallow slope.[1]
2. **Freshwater ecology**: (1) The location of a riparian zone habitats: along the upland and lowland river and stream beds. (2) The ecology around and depending on a marsh, swamp, slough, or estuary.
3. **Fluvial**: A riverbank or stream-bank: the terrain alongside the bed of a river, creek, or stream.
4. **Navigation**: A singular or succession of shoals of alluvium, such as silt and sand, constricting or blocking access to an area. This can be in the course of or at the mouth of a navigable river, in a harbor, or on the continental shelves. An example is a sandbar.

A man-made lake in Keukenhof, Netherlands with grass banks.

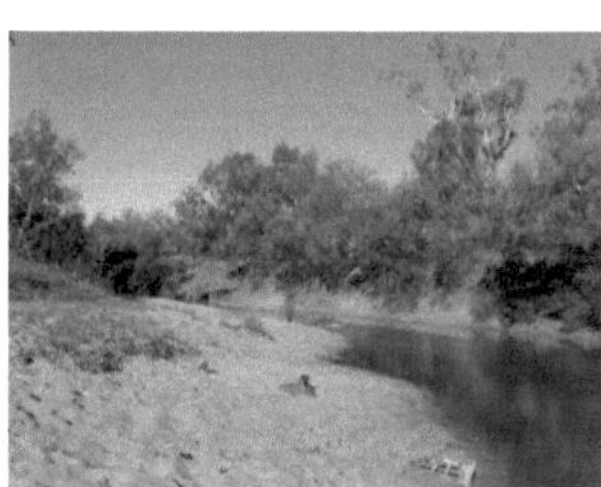

A sloping sandy point bar (close side) and the vegetation-stabilized cut bank (far side) on the Namoi River, New South Wales, Australia. These two constitute the banks of the river.

References

[1] Dictionary for Geographic Words (http://www.docstoc.com/docs/4581771/dictionary-for-geography-words)

- Luna B. Leopold, M. Gordon Wolman, John P. Miller. (1995). *Fluvial processes in geomorphology*. New York: Dover Publications. ISBN 9780486685885.

Lake Henshaw

Lake Henshaw	
Location	San Diego County, California
Coordinates	33°14′30″N 116°45′47″W
Lake type	reservoir
Catchment area	217 square miles (560 km²)
Basin countries	United States
Surface area	1140 acres (460 ha)
Water volume	55000 acre feet (m³)
Shore length[1]	5 miles (8.0 km)
Surface elevation	2723 feet (830 m)
[1] Shore length is not a well-defined measure.	

Lake Henshaw is a Reservoir in San Diego County, California at the southeast base of Palomar Mountain, approximately 70 miles (110 km) miles northeast of San Diego, California and 100 miles (160 km) miles southeast of Los Angeles.[1]

The lake covers approximately 1140 acres (460 ha) and holds 55000 acre feet (m³) of water when full (lowered in 1978 from its original capacity of 203581 acre feet (m³) out of earthquake concerns),[2] in addition to groundwater stored in its local basin. It drains an area of 207 square miles (540 km²) square miles at the source of the San Luis Rey River.

The lake was constructed in 1923 with the building of Henshaw Dam, an earth dam 123 feet (37 m) tall and 650 feet (200 m) long. It is owned by the Vista Irrigation District and used primarily for agricultural irrigation.

References

[1] Gorman, Tom (December 8, 1985). "Agency, Indians Closer to Reaching Agreement Over Water Rights" (http://articles.latimes.com/1985-12-08/local/me-14950_1_water-agencies). .

[2] Zieralski, Ed (October 18, 2003). "Historic lake has suffered through low times, but new concessionaires have high hopes" (http://pqasb.pqarchiver.com/sandiego/access/425485821.html?dids=425485821:425485821&FMT=ABS&FMTS=ABS:FT). San Diego Union-Tribune. .

Llwyn-on Reservoir

Llwyn-on Reservoir	
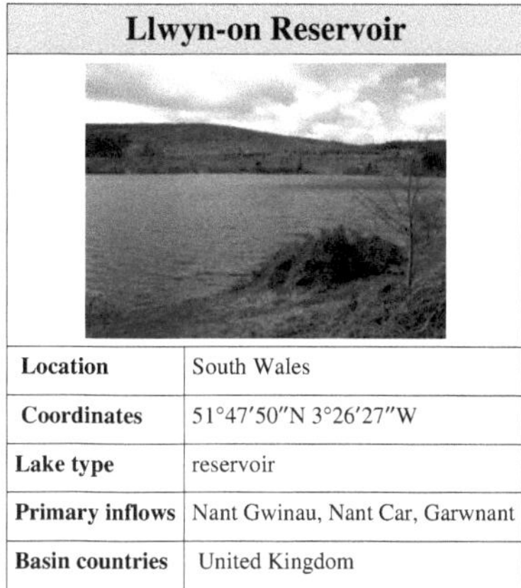	
Location	South Wales
Coordinates	51°47′50″N 3°26′27″W
Lake type	reservoir
Primary inflows	Nant Gwinau, Nant Car, Garwnant
Basin countries	United Kingdom

Llwyn-on Reservoir is the largest and southernmost of the three reservoirs in the Taff Fawr valley in South Wales. It is owned by Welsh Water. It is located in the Brecon Beacons National Park. The eastern half is in the Merthyr Tydfil unitary authority area and the western half is in the Rhondda Cynon Taff unitary authority area. The reservoir is within the historic county boundaries of Breconshire. The dam is adjacent to Llwyn-On village.

Nant Gwinau, Nant Car and Garwnant are the major streams that enter the reservoir.

There are a variety of guided walks and waymarked paths. Environmental sculptures can be found on the Wern and Willow walks. The Taff Trail and the Navvies Line paths link Cefn Coed-y-Cymmer to Brecon. There is one bird hide.

External links

- Welsh Water Recreation Guide [1]PDF (409 KiB)

References

[1] http://www.dwrcymru.com/English/library/publications/community/recreation/Recreation_Guide_2006.pdf

North County, San Diego

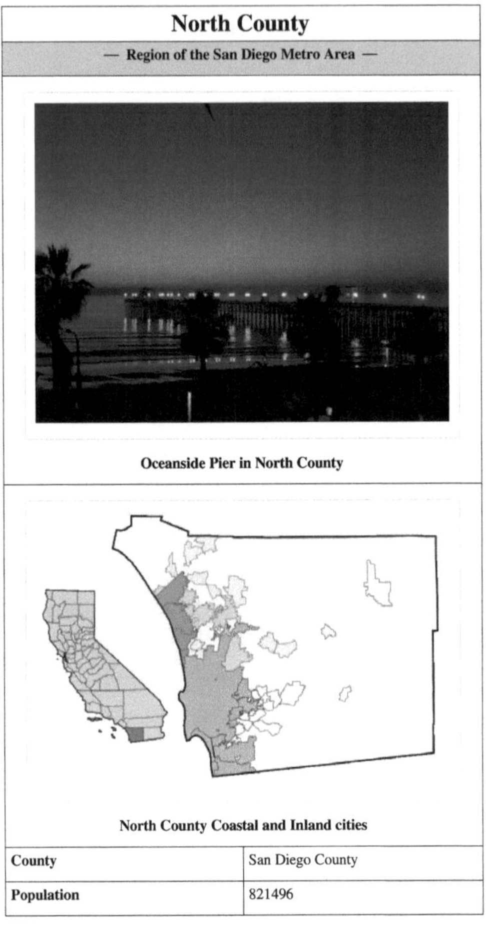

North County

— Region of the San Diego Metro Area —

Oceanside Pier in North County

North County Coastal and Inland cities

County	San Diego County
Population	821496

North County is a region in the northern area of San Diego County, California. It is the second most populous region in the county after San Diego, with an estimated population of 826,985. North County is well known for its affluence, especially in Cardiff-by-the-Sea, Del Mar, Rancho Santa Fe, and Solana Beach where house prices range on average above one millions dollars. [1]

History

The name dates to at least the 1970s, when many of the communities in the area were yet to become incorporated cities and local community decisions were made 40 miles away at the county seat. The North County section of San Diego County has historically been the most expensive region of San Diego, with such affluent neighborhoods as Cardiff-by-the-Sea, Carlsbad, Carmel Valley, Del Mar, Encinitas, Olivenhain, Rancho Santa Fe, and Solana Beach.[2]

In modern times North County continues to grow as a highly influential region of Greater San Diego. The top twenty-five employers in San Diego County are closer to the North County city of Carlsbad than San Diego proper.[3]

Geography

Both coastal and inland North County maintain two types of topography. In North County Coastal the land is generally flat with low rolling hills. The beaches are sandy with occasional tidepools and rocky reefs popping out of the surf. In some cases the coast is dominated by bluff type geography, where the land meeting the ocean sharply drops into the sea with a short beach. In some cases, such as in Encinitas, a whole city can be bisected by a coastal foothill ridge. Mountains soon become imminent as one travels further inland, displaying the rocky peaks of North County Inland. Such peaks included Twin Peaks in Poway, Mount Woodson in Ramona, and Iron Mountain in unincorporated territory between the two cities.

Rivers and creeks flowing west from the mountains farther inland predominantly end up draining into the regions four main lagoons. Throughout their course the way they form many lakes and reservoirs supporting and array of native species.

Definitions vary, but almost always include the communities and cities along Interstate 5 north of Carmel Valley Road and Interstate 15 north of Lake Hodges. For both geographic and political reasons, opinions differ as to whether to include the northern communities within the city limits of San Diego (such as La Jolla, Rancho Bernardo, or Scripps Ranch), or close in communities and cities such as Solana Beach, Cardiff-by-the-Sea, Aviara, Del Mar, Carmel Valley, Rancho Santa Fe, etc.

Regardless, North County is commonly divided into coastal and inland regions. The North County Times, which serves the entire region, shows North County Coastal cities being Oceanside, Carlsbad, Vista, Encinitas, Del Mar, and Solana Beach and North County Inland cities being Escondido, Fallbrook, San Marcos, Poway, Valley Center, Ramona, and Rancho Bernardo.[4]

Ecology

The region has strong ties to its coast. Its notable efforts are popular for preserving many marine environments including lagoons and tidal wetlands, many being last few on the South Coast.[5] Unlike developments in many Orange County coastal cities, the lagoons and large areas of coast have not been so heavily developed. Major lagoons and inlets lining the coast from north to south include: Oceanside Harbor, Buena Vista Lagoon, Agua Hedionda Lagoon, Batiquitos Lagoon and San Elijo Lagoon.

California Least Tern

Flora and fauna

These lagoons provide valuable wetland habitat for many bird, reptile, fish, and plant species. The waters off the coast are also very rich in species diversity; supporting large kelp forests and rocky reefs.

Fish species included the tidewater goby, topsmelt, striped mullet, surfperch and Pacific staghorn sculpin. Leopard sharks forage near the lagoons and pups frequent the shallow rocky reefs off the coast.

Bird species included the Great Blue Heron, Snowy Plover, Clapper Rail and Least tern. The lagoons support various species of shorebirds, wading birds, waterfowl, raptors and diving birds.[5] The number of bird species in the San Dieguito Wetlands have tripled due to restoration projects by Del Mar.[6]

Culture

North County Coastal city of Del Mar

North County contains forty golf courses, including Torrey Pines, which hosted the 2008 U.S. Open.[7] North County is home to Southern California's only five star and five diamond restaurants: Addison at The Grand Del Mar, and El Bizcocho at Rancho Bernardo Inn.[7]

A notable fictional character from North County is Dave Rickards' "Aunt Edna," frequently featured on the popular *Dave, Shelly, and Chainsaw* radio program, which airs in the San Diego area.

Demographics

North County is considered wealthier and more socially conservative than other parts of the metropolitan area such as South Bay. It is the second largest region of the San Diego metropolitan area. There are 325,995 people residing in North County San Diego. 49.1% are males versus 51.9% female. White collar jobs beat blue collar jobs 3:1. More people have B.A. Degrees then AS/AA. degrees 2:1. More people are married then single 2:1.Demographs [8]

Cities

Incorporated cities

North County Coastal city of Oceanside

Populations listed are from the 2010 United States' Census

- Oceanside - 167,086
- Escondido - 143,911
- Carlsbad - 105,328
- Vista - 93,834
- San Marcos - 83,781
- Encinitas - 59,518
- Poway - 47,811
- Solana Beach - 12,867
- Del Mar - 4,161

Unincorporated CDPs

Populations listed are from the 2010 U.S. Census

- Fallbrook - 30,534
- Ramona - 20,292
- Camp Pendleton South - 10,616
- San Diego Country Estates - 10,109
- Valley Center - 9,277

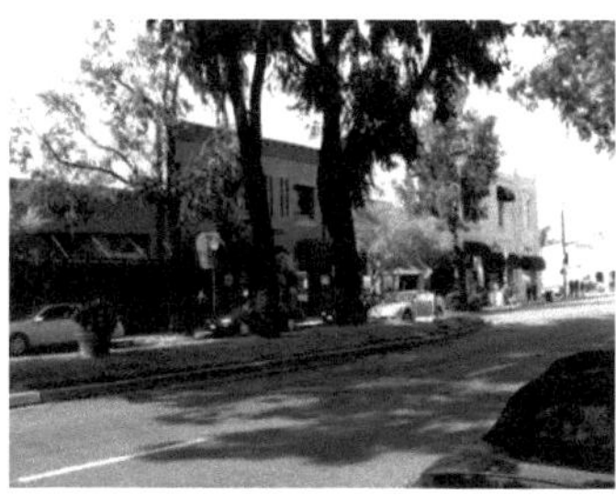

North County Inland city of Escondido

- Camp Pendleton North - 5,200
- Lake San Marcos - 4,437
- Hidden Meadows - 3,485
- Bonsall - 3,982
- Fairbanks Ranch - 3,148
- Rancho Santa Fe - 3,117
- Rainbow - 1,832

Education

Primary and secondary

Colleges and universities

North County is home to several colleges, both traditional four year universities and community colleges. These included California State University, San Marcos (CSUSM) offering undergraduate, graduate, and doctorate programs;[9] and the community colleges Palomar College and MiraCosta College.

Media

Newspaper

The *North County Times* is the main newspaper in the region.

Radio

Landmarks

Important landmarks in North County include Del Mar Racetrack, Lake San Marcos, Oceanside Pier, Stone Steps, Twin Peaks

Del Mar Race Track

Lake San Marcos

Oceanside Pier

Twin Peaks

Politics

Most of North County is located within County Supervisorial District 5, and is represented on the county's Board of Supervisors by Bill Horn. The remainder of North County is in Districts 2 and 3, represented by Dianne Jacob and Pam Slater-Price, respectively.

See also

- East County
- South Bay

References

[1] "San Diego Real Estate Market Reports" (http://www.sandiegorealestatehq.com/north-county-real-estate-report.php). Highland Realty. . Retrieved March 10, 2011.
[2] NPR (http://www.npr.org/templates/story/story.php?storyId=5031130)
[3] California Pacific Airlines. "San Diego North Means Business" (http://www.flycpair.com/landingpage/business.htm). . Retrieved 2011-03-03.
[4] North County Times: Coastal and Inland cities (http://www.nctimes.com/)
[5] "Batiquitos Lagoon" (http://www.batiquitosfoundation.org/newsite/index.php). Batiquitos Lagoon Foundation. . Retrieved 2011-03-03.
[6] "DEL MAR:Restoring the wetlands" (http://www.nctimes.com/news/local/del-mar/article_64da19e1-65cc-5a4f-90ad-dd85422905e2.html). North County Times. . Retrieved March 10, 2011.
[7] San Diego's North County (http://www.sandiego.org/article/Visitors/119)
[8] http://homes.point2.com/Neighborhood/US/California/San-Diego-County/San-Diego/North-San-Diego-County-Demographics.aspx
[9] "CSUSM: CSU San Marcos Academics" (http://www.csusm.edu/academics/). California State University San Marcos. . Retrieved June 12, 2011.

External links

- San Diego North Convention & Visitors Bureau (http://www.sandiegonorth.com/)
- City of Oceanside (http://www.ci.oceanside.ca.us/)
- City of Escondido (http://www.escondido.org/)
- City of Carlsbad (http://www.carlsbadca.gov/Pages/default.aspx)
- City of Vista (http://www.cityofvista.com/)
- City of San Marcos (http://www.ci.san-marcos.ca.us/)
- City of Encinitas (http://www.cityofencinitas.org/)
- City of Poway (http://www.poway.org/index.aspx?page=1)
- City of Solana Beach (http://www.ci.solana-beach.ca.us/csite/cms/home.htm)
- City of Del Mar (http://www.delmar.ca.us/default.aspx)

Palomar Mountain

Palomar Mountain	
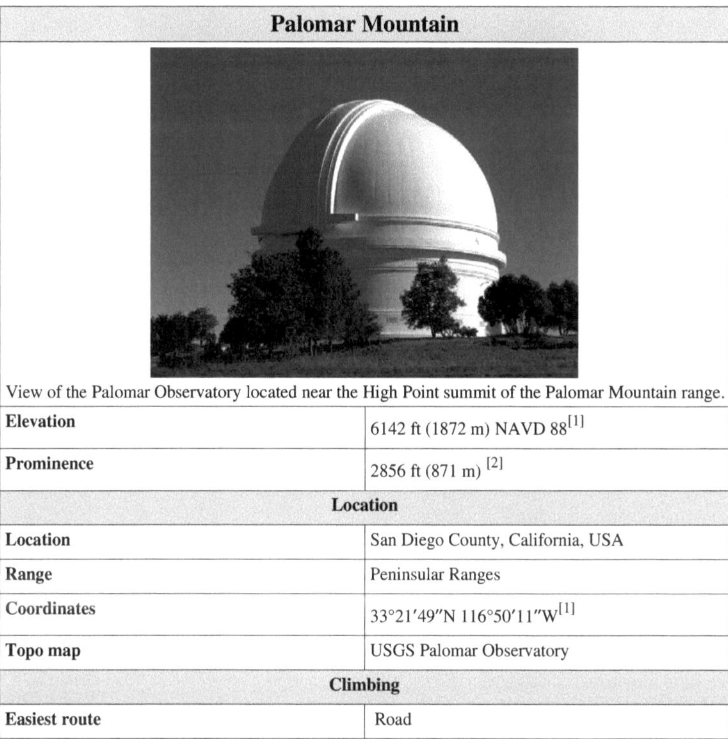 View of the Palomar Observatory located near the High Point summit of the Palomar Mountain range.	
Elevation	6142 ft (1872 m) NAVD 88[1]
Prominence	2856 ft (871 m) [2]
Location	
Location	San Diego County, California, USA
Range	Peninsular Ranges
Coordinates	33°21′49″N 116°50′11″W[1]
Topo map	USGS Palomar Observatory
Climbing	
Easiest route	Road

Palomar Mountain is a mountain in the Peninsular Ranges in northern San Diego County, southern California, United States. It is famous as the location of the Palomar Observatory and Hale Telescope, and known for the Palomar Mountain State Park.

History

The Spanish name "Palomar", in English meaning "pigeon roost," comes from Spanish colonial era in Alta California when Palomar Mountain was known as the home of Band-tailed Pigeons.[3]

During the 1890s, the human population was sufficient to support three public schools, and it was a popular summer resort for Southern California, with three hotels in operation part of the time, and a tent city in Doane Valley each summer.

Palomar Observatory

Palomar Mountain is most famous as being home since 1936 to the Palomar Observatory, and the giant Hale Telescope. The 200-inch telescope was the world's largest and most important telescope from 1949 until 1992. The observatory currently consists of three large telescopes.

Palomar Mountain State Park

Palomar Mountain is the location of Palomar Mountain State Park, a California State Park. There are campgrounds for vacationers, as well as a campground for local school children. The park averages 70,000 visitors annually. High Point in the Palomar Mountain range is one of the highest peaks in San Diego County, at 6140 feet (1871 m), although it is still dwarfed by the higher 11500 feet (3505 m) San Bernardino Mountains a relatively short distance north in San Bernardino County and Riverside County and the 14500 feet (4420 m) high Mount Whitney some 250 mi (402 km) further north.

At the base of Palomar Mountain on S6 is Oak Knoll Campground, formerly known as Palomar Gardens. Palomar Gardens was made somewhat famous by an earlier resident George Adamski. Adamski had an observatory at Palomar Gardens and photographed objects in the night sky that he claimed were UFOs. Adamski co-authored *Flying Saucers Have Landed* in 1953,[4] about his alien encounter experiences. The 1977 film *The Crater Lake Monster* had many scenes filmed on Palomar Mountain, including scenes shot at the summit restaurant, but not the scenes of the monster in a lake.[5]

Doane Valley, located within the State Park, is home to the Camp Palomar Outdoor School for 6th grade students in the San Diego Unified School District.[6]

Palomar Mountain State Park is one of 70 California State Parks scheduled to close due to budget cuts. The park will close July 2012 to achieve part of the $11 million dollar budget reduction for the 2011/2012 fiscal year.[7]

Access

South Grade Road, the stretch of San Diego County Route S6 going from State Route 76 to the summit, is popular among motorcycle riders and sports car drivers due to its challenging nature [8] (over 20 hairpin turns over the distance of less than 7 mi (11 km)). According to fire department records, there have been 26 reported motorcycle injury accidents on the mountain in 2005. In 2004, the figure was 23. In 2003 there were 26.[8] The Luiseno Indian name for Palomar Mountain was "Paauw" and High Point was called "Wikyo."[9]

See also

- Banana slug

References

[1] "Palomar" (http://www.ngs.noaa.gov/cgi-bin/ds_mark.prl?PidBox=DX5064). *NGS data sheet*. U.S. National Geodetic Survey. . Retrieved 2009-08-03.

[2] "Palomar Mountain, California" (http://www.peakbagger.com/peak.aspx?pid=1452). Peakbagger.com. . Retrieved 2009-08-07.

[3] Wood, Catherine M. (1937). *Palomar from teepee to telescope* (http://www.peterbrueggeman.com/palomarhistory/wood.pdf). San Diego: Frye & Smith. . Retrieved 2009-10-31.

[4] Leslie, Desmond; George Adamski (1953). *Flying saucers have landed*. New York: British Book Centre. ISBN 0854351809.

[5] "The Crater Lake Monster" (http://www.imdb.com/title/tt0075888/). Crown International Pictures. . Retrieved 2009-09-23.

[6] "Camp Palomar Outdoor School – Directions" (http://www.sandi.net/204510112512039907/blank/browse.asp?A=383&BMDRN=2000&BCOB=0&C=56179). San Diego Unified School District. . Retrieved 2009-08-07.

[7] "California State Park Closures Announced" (http://roughin.it/2011/05/california-state-park-closures-announced/). Roughin.It. . Retrieved 25 May 2011.

[8] J. Harry Jones (September 25, 2005). "Twists, turns, trouble" (http://legacy.signonsandiego.com/news/northcounty/20050925-9999-2m25moto.html). The San Diego Union-Tribune. . Retrieved 2010-01-28.

[9] Sparkman, Philip Stedman (1908). *The Culture of the Luiseño Indians* (http://www.peterbrueggeman.com/palomarhistory/sparkman_luiseno.pdf). Berkeley: University of California Press. . Retrieved 2009-09-27.

Sources

- "High Point" (http://geonames.usgs.gov/pls/gnispublic/f?p=gnispq:3:::NO::P3_FID:271572). Geographic Names Information System, U.S. Geological Survey. Retrieved 2009-08-07.
- "Palomar Mountain East Grade (S7)" (http://www.sundaymorningrides.com/road/8716970/). SundayMorningRides.com. Retrieved 2011-05-09.
- "Palomar Mountain South Grade (S6)" (http://www.sundaymorningrides.com/road/9367624/). SundayMorningRides.com. Retrieved 2011-05-09.

External links

- "Palomar Mountain State Park" (http://www.parks.ca.gov/default.asp?page_id=637). California State Parks. Retrieved 2011-05-09.
- "Palomar Mountain State Park" (http://www.palomarsp.org/index.htm). Cuyamaca Rancho State Park Interpretive Association. Retrieved 2011-09-05.
- "Palomar Mountain History Resources" (http://www.peterbrueggeman.com/palomarhistory/). Peter Brueggeman. Retrieved 2011-05-09.
- "Palomar Mountain" (http://www.borregospringschamber.com/abdsp/cdd/palomar.htm). Borrego Springs Chamber of Commerce. Retrieved 2011-09-05.

San Diego

San Diego
— City —
City of San Diego
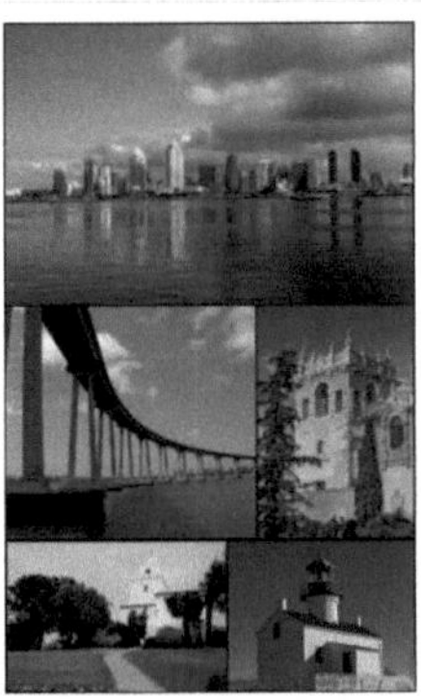 Images from top, left to right: San Diego Skyline, Coronado Bridge, museum in Balboa Park, Serra Museum in Presidio Park and the Old Point Loma Lighthouse
 **Flag**  **Seal**
Nickname(s): America's Finest City
Motto: *Semper Vigilans* (Latin for "Ever Vigilant")
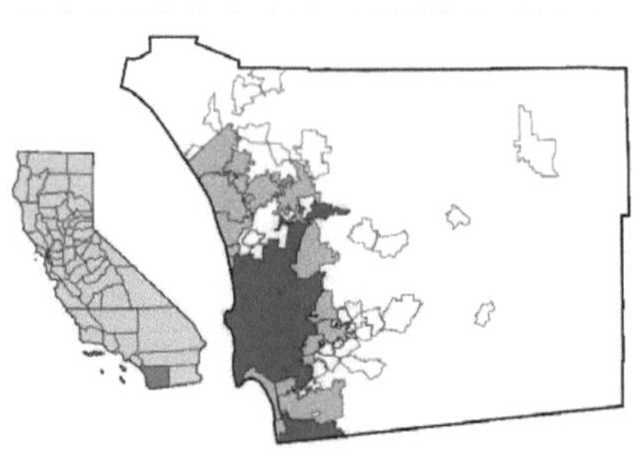 Location of San Diego within San Diego County

San Diego	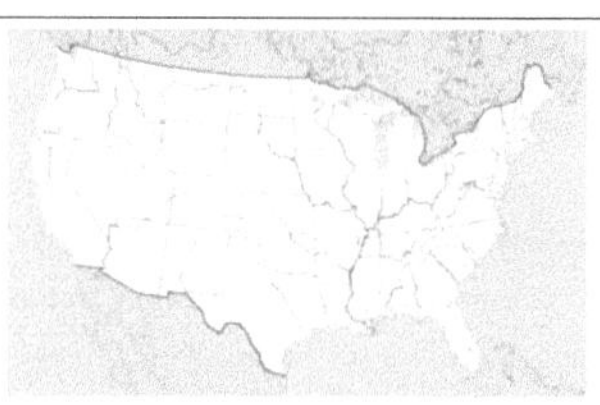 Location in the United States
Coordinates: 32°42′54″N 117°09′45″W	
Country	United States
State	California
County	San Diego
Founded	July 16, 1769
Incorporated	March 27, 1850
Government	
- Type	Mayor-council
- Body	San Diego City Council
- Mayor	Jerry Sanders
- City Attorney	Jan Goldsmith
- City Council members	
Area[1]	
- City	372.398 sq mi (964.506 km^2)
- Land	325.188 sq mi (842.233 km^2)
- Water	47.210 sq mi (122.273 km^2) 12.68%
Elevation	72–400 ft (22 m)
Population (Census 2010)	
- City	1301617
- Rank	1st in San Diego County 2nd in California 8th in the United States
- Density	4002.6/sq mi (1545.4/km^2)
- Metro	3,095,313
Demonym	San Diegan
Time zone	PST (UTC-8)
- Summer (DST)	PDT (UTC-7)
ZIP code	92101-92117, 92119-92124, 92126-92140, 92142, 92145, 92147, 92149-92155, 92158-92172, 92174-92177, 92179, 92182, 92184, 92186, 92187, 92190-92199
Area code(s)	619, 858
FIPS code	66000

GNIS feature ID	1661377
Website	www.sandiego.gov [2]

San Diego /ˌsændiːˈeɪɡoʊ/ is the eighth-largest city in the United States and second-largest city in California. The city is located on the coast of the Pacific Ocean in Southern California, immediately adjacent to the Mexican border. The birthplace of California,[3] San Diego is known for its mild year-round climate, its natural deep-water harbor, and its long association with the U.S. Navy. The population was 1,301,617 at the 2010 census.[4]

Historically home to the Kumeyaay people, San Diego was the first site visited by Europeans on what is now the West Coast of the United States. Upon landing in San Diego Bay in 1542, Juan Cabrillo claimed the entire area for Spain, forming the basis for the settlement of Alta California 200 years later. The Presidio and Mission of San Diego, founded in 1769, were the first European settlement in what is now California. In 1821, San Diego became part of newly independent Mexico, and in 1850, became part of the United States following the Mexican-American War and the admission of California to the union.

The city is the county seat of San Diego County and is the economic center of the San Diego–Carlsbad–San Marcos metropolitan area as well as the San Diego–Tijuana metropolitan area. San Diego's main economic engines are military and defense-related activities, tourism, international trade, and manufacturing. The presence of the University of California, San Diego (UCSD), with the affiliated UCSD Medical Center, has helped make the area a center of research in biotechnology.

History

Kumeyaay people lived in San Diego for more than 10,000 years before Europeans settled there.

The area of San Diego has been inhabited for more than 10,000 years by the Kumeyaay people.[5] The first European to visit the region was Portuguese-born explorer Juan Rodríguez Cabrillo sailing under the flag of Castile. Sailing his flagship *San Salvador* from Navidad, New Spain, Cabrillo claimed the bay for the Spanish Empire in 1542 and named the site 'San Miguel'.[6] In November 1602, Sebastián Vizcaíno was sent to map the California coast. Arriving on his flagship *San Diego*, Vizcaíno surveyed the harbor and what are now Mission Bay and Point Loma and named the area for the Catholic Saint Didacus, a Spaniard more commonly known as *San Diego de Alcalá*. On November 12, 1602, the first Christian religious service of record in Alta California was conducted by Friar Antonio de la Ascensión, a member of Vizcaíno's expedition, to celebrate the feast day of San Diego.[7]

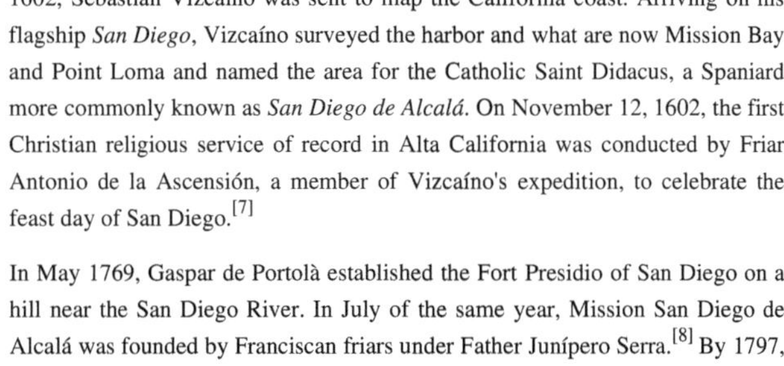

In May 1769, Gaspar de Portolà established the Fort Presidio of San Diego on a hill near the San Diego River. In July of the same year, Mission San Diego de Alcalá was founded by Franciscan friars under Father Junípero Serra.[8] By 1797, the mission boasted the largest native population in Alta California, with over 1,400 neophytes living in and around the mission proper.[9] Mission San Diego was the southern anchor in California of the historic mission trail El Camino Real. Both the Presidio and the Mission are National Historic Landmarks.[10] [11]

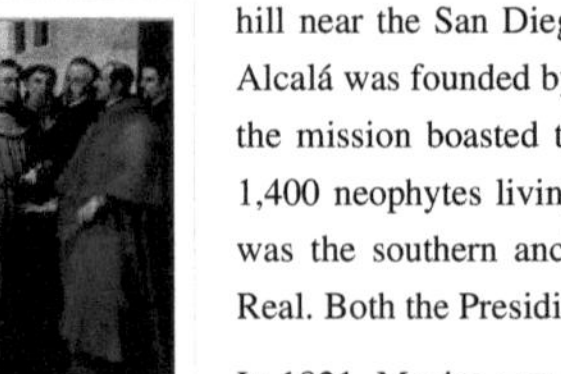

Namesake of the city, Didacus of Alcalá: *Saint Didacus in Ecstasy Before the Cross* by Murillo (Musée des Augustins)

In 1821, Mexico won its independence from Spain, and San Diego became part of the Mexican state of Alta California. The fort on Presidio Hill was gradually abandoned, while the town of San Diego grew up on the level land below Presidio Hill. The Mission was secularized by the Mexican government, and most of the Mission lands were distributed to wealthy Californio settlers.

Mission San Diego de Alcalá

As a result of the Mexican-American War of 1846–1848, the territory of Alta California, including San Diego, was ceded to the United States by Mexico. The Battle of San Pasqual, a battle of the Mexican-American War, was fought in 1846 in the San Pasqual Valley which is now part of the city of San Diego. The state of California was admitted to the United States in 1850. That same year San Diego was designated the seat of the newly established San Diego County and was incorporated as a city. The initial city charter was established in 1889 and today's city charter was adopted in 1931.[12]

Namesake of Horton Plaza, Alonzo Horton developed "New Town" which became Downtown San Diego.

The original town of San Diego was located at the foot of Presidio Hill, in the area which is now Old Town San Diego State Historic Park. The location was not ideal, being several miles away from navigable water. In the late 1860s, Alonzo Horton promoted a move to "New Town", several miles south of the original settlement, in the area which became Downtown San Diego. People and businesses flocked to New Town because of its location on San Diego Bay convenient to shipping. New Town quickly eclipsed the original settlement, known to this day as Old Town, and became the economic and governmental heart of the city.[13]

In the early part of the 20th century, San Diego hosted two World's Fairs: the Panama-California Exposition in 1915 and the California Pacific International Exposition in 1935. Both expositions were held in Balboa Park, and many of the Spanish/Baroque-style buildings that were built for those expositions remain to this day as central features of the park. The buildings were intended to be temporary structures, but most remained in continuous use until they progressively fell into disrepair. Most were eventually rebuilt, using castings of the original facades to retain the architectural style.[14] The menagerie of exotic animals featured at the 1915 exposition provided the basis for the San Diego Zoo.[15]

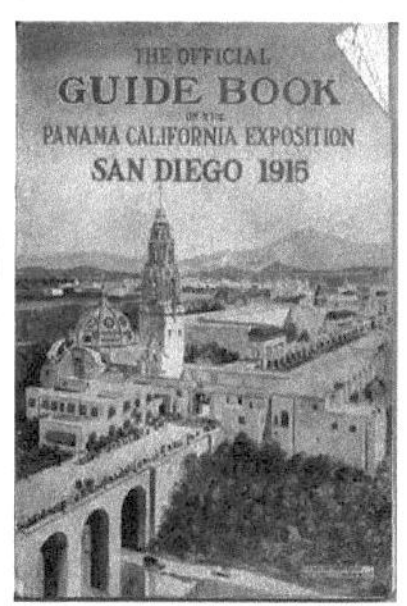

Balboa Park on the cover of a guidebook for the World Exposition of 1915

Significant U.S. Navy presence began in 1901 with the establishment of the Navy Coaling Station in Point Loma, and expanded greatly during the 1920s.[16] By 1930 the city was host to Naval Base San Diego, Naval Training Center San Diego, San Diego Naval Hospital, Camp Matthews, and Camp Kearny (now Marine Corps Air Station Miramar). The city was also an early center for aviation: as early as World War I San Diego was proclaiming itself "The Air Capital of the West."[17] The city was home to important airplane developers and manufacturers like Ryan Airlines (later Ryan Aeronautical), founded in 1925, and Consolidated Aircraft (later Convair), founded in 1923. Charles A. Lindbergh's plane The Spirit of St. Louis was built in San Diego in 1927 by Ryan Airlines.[17]

During World War II, San Diego became a major hub of military and defense activity, due to the presence of so many military installations and defense manufacturers. The city's population grew rapidly during and after World War II, more than doubling between 1930 (147,995) and 1950 (333,865).[18] After World War II, the military continued to play a major role in the local economy, but post-Cold War cutbacks took a heavy toll on the local defense and aerospace industries. The resulting downturn led San Diego leaders to seek to diversify the city's economy by focusing on research and science, as well as tourism.[19]

Downtown San Diego was in decline in the 1960s and 1970s but experienced some urban renewal since the early 1980s, including the opening of Horton Plaza, the revival of the Gaslamp Quarter, and the construction of the San

Diego Convention Center; Petco Park opened in 2004.[20]

Geography

The San Diego-Tijuana metropolitan area

The city of San Diego lies on deep canyons and hills separating its mesas, creating small pockets of natural parkland scattered throughout the city and giving it a hilly geography. Traditionally, San Diegans have built their homes and businesses on the mesas, while leaving the canyons relatively wild.[21] Thus, the canyons give parts of the city a segmented feel, creating gaps between otherwise proximate neighborhoods and contributing to a low-density, car-centered environment. The San Diego River runs through the middle of San Diego from east to west, creating a river valley which serves to divide the city into northern and southern segments. Several reservoirs and Mission Trails Regional Park also lie between and separate developed areas of the city.

Notable peaks within the city limits include Cowles Mountain, the highest point in the city at 1593 feet (486 m); Black Mountain at 1558 feet (475 m); and Mount Soledad at 824 feet (251 m). The Cuyamaca Mountains and Laguna Mountains rise to the east of the city, and beyond the mountains are desert areas. The Cleveland National Forest is a half-hour drive from downtown San Diego. Numerous farms are found in the valleys northeast and southeast of the city.

Communities and neighborhoods

The city of San Diego recognizes 52 individual areas as Community Planning Areas.[22] Within a given planning area there may be several distinct neighborhoods. Altogether the city contains more than 100 identified neighborhoods.[23]

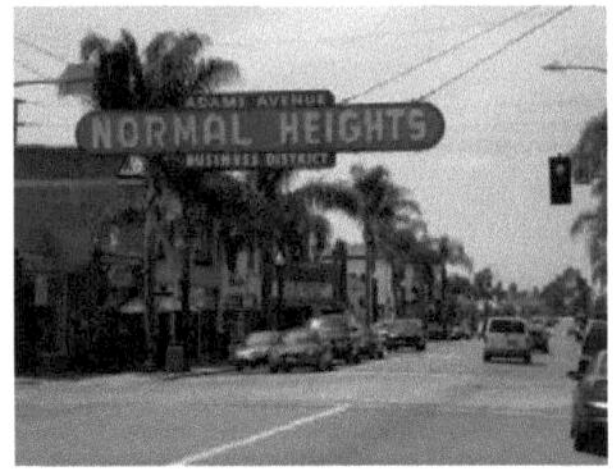

Normal Heights, a neighborhood

Downtown San Diego is located on San Diego Bay. Balboa Park encompasses several mesas and canyons to the northeast, surrounded by older, dense urban communities including Hillcrest and North Park. To the east and southeast lie City Heights, the College Area, and Southeast San Diego. To the north lies Mission Valley and Interstate 8. The communities north of the valley and freeway, and south of Marine Corps Air Station Miramar, include Clairemont, Kearny Mesa, Tierrasanta, and Navajo. Stretching north from Miramar are the northern suburbs of Mira Mesa, Scripps Ranch, Rancho Peñasquitos, and Rancho Bernardo. The far northeast portion of the city encompasses Lake Hodges and the San Pasqual Valley, which holds an agricultural preserve. Carmel Valley and Del Mar Heights occupy the northwest corner of the city. To their south are Torrey Pines State Reserve and the business center of the Golden Triangle. Further south are the beach and coastal communities of La Jolla, Pacific Beach, and Ocean Beach. Point Loma occupies the peninsula across San Diego Bay from downtown. The communities of South San Diego, such as San Ysidro and Otay Mesa, are located next to the Mexican border, and are physically separated from the rest of the city by the cities of National City and Chula Vista; a narrow strip of land at the bottom of San Diego Bay connects these southern neighborhoods with the rest of the city.

Unlike some areas of Southern California, where neighborhoods and even cities may run into each other without a clear demarcation, San Diego neighborhood boundaries tend to be clearly understood by their residents based on geographical boundaries like canyons and street patterns.[24] The city recognized the importance of its neighborhoods when it organized its 2008 General Plan around the concept of a "City of Villages".[25]

Cityscape

San Diego was originally centered in the Old Town district, but by the late 1860s the center of focus had relocated to the bayfront in the belief that this new location would increase trade. As the "New Town" – present-day Downtown – waterfront location quickly developed, it eclipsed Old Town as the center of San Diego.[26]

The development of skyscrapers over 300 feet (91 m) in San Diego is attributed to the construction of the El Cortez Apartment Hotel in 1927, the tallest building in the city from 1927 to 1963.[27] As time went on multiple buildings claimed the title of San Diego's tallest skyscraper, including the Union Bank of California Building and Symphony Towers. Currently the tallest building in San Diego is One America Plaza, standing 500 feet (150 m) tall, which was completed in 1991.[28] The downtown skyline contains no super-talls, as a regulation put in place by the Federal Aviation Administration in the 1970s set a 500 feet (152 m) limit on the height of buildings due to the proximity of San Diego International Airport.[29] An iconic description of the skyline includes its skyscrapers being compared to the tools of a toolbox.[30] Within the city limits are multiple skylines composed of high-rises and mid-rises, including University City, Rancho Bernardo, Centre City, Carmel Valley and La Jolla Village.

Panorama of San Diego as viewed from North Island

Climate

A surfer at Black's Beach

San Diego is one of the top-ten best climates in the *Farmer's Almanac*[31] and is one of the two best summer climates in America as scored by The Weather Channel.[32] Under the Köppen climate classification system, the San Diego area straddles areas of Mediterranean climate (CSa) to the north and Semi-arid climate (BSh) to the south and east.[33] As a result, it is often described as "arid Mediterranean" and "Semi-arid Steppe". San Diego's climate is characterized by warm, dry summers and mild winters with most of the annual precipitation falling between December and March. The city has mild, mostly dry weather, with an average of 201 days above 70 °F (21 °C) and low rainfall (9–13 inches [23–33 cm] annually).

The climate in the San Diego area, like much of California, often varies significantly over short geographical distances resulting in microclimates. In San Diego's case this is mainly due to the city's topography (the Bay, and the numerous hills, mountains, and canyons). Frequently, particularly during the "May gray/June gloom" period, a thick "marine layer" cloud cover will keep the air cool and damp within a few miles of the coast, but will yield to bright cloudless sunshine approximately 5–10 miles (8.0–16 km) inland.[34] Sometimes the June gloom can last into July, causing cloudy skies over most of San Diego for the entire day.[35] [36] Even in the absence of June gloom, inland areas tend to experience much more significant temperature variations than coastal areas, where the ocean serves as a moderating influence. Thus, for example, downtown San Diego averages January lows of 50 °F (10 °C) and August highs of 78 °F (26 °C). The city of El Cajon, just 10 miles (16 km) inland from downtown San Diego, averages January lows of 42 °F (6 °C) and August highs of 88 °F (31 °C).

A sign of global warming, scientists at Scripps Institution of Oceanography say the average surface temperature of the water at Scripps Pier in the California Current has increased by almost 3 degrees since 1950.[37]

Surfers in Pacific Beach

Rainfall along the coast averages about 10 inches (250 mm) of precipitation annually. The average (mean) rainfall is 10.65 inches (271 mm) and the median is 9.6 inches (240 mm).[38] Most of the rainfall occurs during the cooler months. The months of December through March supply most of the rain, with February the only month averaging 2 inches (51 mm) or more of rain. The months of May through September tend to be almost completely dry. Though there are few wet days per month during the rainy period, rainfall can be heavy when it does fall. Rainfall is usually greater in the higher elevations of San Diego; some of the higher elevation areas of San Diego can receive 11–15 inches (280–380 mm) of rain a year.

Snow in the city is so rare that it has been observed only five times in the century-and-a-half that records have been kept. In 1949 and 1967, snow stayed on the ground for a few hours in higher locations like Point Loma and La Jolla. The other three occasions, in 1882, 1946, and 1987, involved flurries but no accumulation.[39]

Official temperature record-keeping began in San Diego in 1872,[40] although other weather records go back further. The city's first official weather station was located at Mission San Diego from 1849 to 1858. From August 1858 until 1940, the official weather station was located at a series of downtown buildings, and the station has been at Lindbergh Field since February 1940.[41]

Ecology

Coastal canyon in Torrey Pines State Reserve

Like most of southern California, the majority of San Diego's current area was originally occupied by chaparral, a plant community made up mostly of drought-resistant shrubs. The endangered Torrey Pine has the bulk of its population in San Diego in a stretch of protected chaparral along the coast. The steep and varied topography and proximity to the ocean create a number of different habitats within the city limits, including tidal marsh and canyons. The chaparral and coastal sage scrub habitats in low elevations along the coast are prone to wildfire, and the rates of fire have increased in the 20th century, due primarily to fires starting near the borders of urban and wild areas.[44]

San Diego's broad city limits encompass a number of large nature preserves, including Torrey Pines State Reserve, Los Peñasquitos Canyon Preserve, and Mission Trails Regional Park. Torrey Pines State Reserve and a coastal strip continuing to the north constitute the only location where the rare species of Torrey Pine, *P. torreyana torreyana*, is found.[45]

San Diego viewed against the Witch Creek Fire smoke

Due to the steep topography that prevents or discourages building, along with some efforts for preservation, there are also a large number of canyons within the city limits that serve as nature preserves, including Switzer Canyon, Tecolote Canyon Natural Park,[46] and Marian Bear Memorial Park in the San Clemente Canyon,[47] as well as a number of small parks and preserves.

San Diego County has one of the highest counts of animal and plant species that appear on the endangered species list among counties in the United States.[48] Because of its diversity of habitat and its position on the Pacific Flyway, San Diego County has recorded the presence of 492 bird species, more than any other region in the country.[49] San Diego always scores very high in the number of bird species observed in the annual Christmas Bird Count, sponsored by the Audubon Society, and it is known as one of the "birdiest" areas in the United States.[50][51]

San Diego and its backcountry are subject to periodic wildfires. In October 2003, San Diego was the site of the Cedar Fire, which has been called the largest wildfire in California over the past century.[52] The fire burned 280000 acres (1100 km^2), killed 15 people, and destroyed more than 2,200 homes.[53] In addition to damage caused by the fire, smoke resulted in a significant increase in emergency room visits due to asthma, respiratory problems, eye irritation, and smoke inhalation; the poor air quality caused San Diego County schools to close for a week.[54] Wildfires four years later destroyed some areas, particularly within the communities of Rancho Bernardo, Rancho Santa Fe, and Ramona.[55]

Demographics

The city had a population of 1,307,402 in 2010, according to the census that year, on a land area of 372.1 square miles (963.7 km^2). The urban area of San Diego extends beyond the administrative city limits and had a total 2010 population of 2,880,000, making it the third-largest urban area in California.

Historical populations		
Census	**Pop.**	**%±**
1850	500	—
1860	731	46.2%
1870	2300	214.6%
1880	2637	14.7%
1890	16159	512.8%
1900	17700	9.5%
1910	39578	123.6%
1920	74361	87.9%
1930	147995	99.0%
1940	203341	37.4%
1950	333865	64.2%
1960	573224	71.7%
1970	696769	21.6%
1980	875538	25.7%

1990	1110549	26.8%
2000	1223400	10.2%
2010	1307402	6.9%
source:[18] [56]		

As of the Census of 2010, there were 1,307,402 people living in the city of San Diego.[57] That represents a population increase of just under 7% from the 1,223,400 people, 450,691 households, and 271,315 families reported in 2000.[58] The estimated city population in 2009 was 1,306,300. The population density was 3,771.9 people per square mile (1,456.4/km^2). The racial makeup of San Diego was 58.9% White, 6.7% African American, 0.6% Native American, 15.9% Asian (5.9% Filipino, 2.7% Chinese, 2.5% Vietnamese, 1.3% Indian, 1.0% Korean, 0.7% Japanese, 0.4% Laotian, 0.3% Cambodian, 0.1% Thai). 0.5% Pacific Islander, 12.3% from other races, and 5.1% from two or more races. Hispanic or Latino of any race were 28.8%.[59] [60] Among the Hispanic population, 24.9% are Mexican, and 0.6% are Puerto Rican.

A U.S. Navy vice admiral and an intelligence specialist celebrating Hispanic American Heritage Month in San Diego

As of January 1, 2008 estimates by the San Diego Association of Governments revealed that the household median income for San Diego rose to $66,715, up from $45,733, and that the city population rose to 1,336,865, up 9.3% from 2000.[61] The population was 45.3% non-Hispanic whites, 27.7% Hispanics, 15.6% Asians/Pacific Islanders, 7.1% blacks, 0.4% American Indians, and 3.9% from other races. Median age of Hispanics was 27.5 years, compared to 35.1 years overall and 41.6 years among non-Hispanic whites; Hispanics were the largest group in all ages under 18, and non-Hispanic whites constituted 63.1% of population 55 and older.

In 2000 there were 451,126 households out of which 30.2% had children under the age of 18 living with them, 44.6% were married couples living together, 11.4% had a female householder with no husband present, and 39.8% were non-families. Households made up of individuals account for 28.0% and 7.4% had someone living alone who was 65 years of age or older. The average household size was 2.61 and the average family size was 3.30.

The U.S. Census Bureau reported that in 2000, 24.0% of San Diego residents were under 18, and 10.5% were 65 and over.[62] The median age was 32; two-thirds of the population was under 35.[63] The San Diego County regional planning agency, SANDAG, provides tables and graphs breaking down the city population into 5-year age groups.[64] In 2000, the median income for a household in the city was $45,733, and the median income for a family was $53,060.[65] Males had a median income of $36,984 versus $31,076 for females. The per capita income for the city was $23,609.[65] According to *Forbes* in 2005, San Diego was the fifth wealthiest U.S. city[66] but about 10.6% of families and 14.6% of the population were below the poverty line, including 20.0% of those under age 18 and 7.6% of those age 65 or over.[65] Nonetheless, San Diego was rated the fifth-best place to live in the United States in 2006 by *Money* magazine.[67]

Crime

According to *Forbes* magazine, San Diego was the ninth-safest city in the top 10 list of safest cities in the U.S. in 2010.[68] Like most major cities, San Diego had a declining crime rate from 1990 to 2000. Crime slightly increased in the early 2000s.[69] [70] [71] In 2004, San Diego had the sixth lowest crime rate of any U.S. city with over half a million residents.[71] From 2002 to 2006, the crime rate overall dropped 0.8%, though not evenly by category. While violent crime decreased 12.4% during this period, property crime increased 1.1%. Total property crimes per 100,000 people were lower than the national average in 2008.[72]

Economy

The largest sectors of San Diego's economy are defense/military, tourism, international trade, and research/manufacturing, respectively.[73] [74]

Defense and military

Space and Naval Warfare Systems Command (SPAWAR)

The economy of San Diego is influenced by its deepwater port, which includes the only major submarine and shipbuilding yards on the West Coast. Several major national defense contractors were started and are headquartered in San Diego, including General Atomics, Cubic, and NASSCO.

San Diego hosts the largest naval fleet in the world:[75] it was in 2008 was home to 53 ships, over 120 tenant commands, and more than 35,000 sailors, soldiers, Department of Defense civilian employees and contractors.[76] About 5 percent of all civilian jobs in the county are military-related, and 15,000 businesses in San Diego County rely on Department of Defense contracts.[76]

F/A-18 Hornet flying over San Diego

Military bases in San Diego include US Navy facilities, Marine Corps bases, and Coast Guard stations. Marine Corps institutions in the city of San Diego include Marine Corps Air Station Miramar and Marine Corps Recruit Depot San Diego. The Navy has several institutions in the city, including Naval Base Point Loma, Naval Base San Diego (also known as the 32nd Street Naval Station), Bob Wilson Naval Hospital, the Space and Naval Warfare Systems Center San Diego and Space and Naval Warfare Systems Command. Also near San Diego but not within the city limits are Naval Amphibious Base Coronado and Naval Air Station North Island (which operates Naval Auxiliary Landing Facility San Clemente Island, Silver Strand Training Complex, and the Outlying Field Imperial Beach). San Diego is known as the "birthplace of naval aviation".[77]

The city is "home to the majority of the U.S. Pacific Fleet's surface combatants, all of the Navy's West Coast amphibious ships and a variety of Coast Guard and Military Sealift Command vessels".[76] One Nimitz class supercarrier, (the USS *Carl Vinson*),[78] five amphibious assault ships, several *Los Angeles*-class "fast attack" submarines, the Hospital Ship USNS *Mercy*, carrier and submarine tenders, destroyers, cruisers, frigates, and many smaller ships are home-ported there. Four Navy vessels have been named USS *San Diego*.[79]

Tourism

Tourism is a major industry owing to the city's climate, its beaches, and numerous tourist attractions such as Balboa Park, Belmont amusement park, San Diego Zoo, San Diego Zoo Safari Park, and SeaWorld San Diego. San Diego's Spanish and Mexican heritage is reflected in the many historic sites across the city, such as Mission San Diego de Alcala and Old Town San Diego State Historic Park. Annual events in San Diego include Comic-Con, the Farmers Insurance Open golf tournament, the San Diego Black Film Festival, and Street Scene Music Festival. Transient Occupancy Taxes (TOT) create funding for the City of San Diego Commission for Arts and Culture.[80]

San Diego County hosted more than 30 million visitors in 2009, of whom approximately half stayed overnight and half were day visitors; collectively they spent an estimated $15 billion locally.[81] The San Diego Convention Center hosted 68 out-of-town conventions and trade shows in 2009, attracting more than 600,000 visitors.[81]

San Diego's cruise ship industry is the second largest in California; each cruise ship call injects an estimated $2 million (from the purchase of food, fuel, supplies, and maintenance services) into the local economy.[82]

Numerous cruise lines, including Celebrity, Crystal and Princess, operate out of San Diego. However, cruise ship business has been in steady decline since peaking in 2008, when the Port hosted over 250 ship calls and more than 900,000 passengers. By 2011 the number of ship calls had fallen to 103 (estimated).[83] Holland America and Carnival Cruises had operated weekly cruises to the Mexican Riviera, but announced that they will no longer do so after April 2012, an economic loss to the region of more than $100 million.[83] The decline is blamed on the slumping economy as well as fear of travel to Mexico due to well-publicized violence there.[84]

International trade

San Diego's commercial port and its location on the United States-Mexico border make international trade an important factor in the city's economy. The city is authorized by the United States government to operate as a Foreign Trade Zone.[85]

The city shares a 15-mile (24 km) border with Mexico that includes two border crossings. San Diego hosts the busiest international border crossing in the world, in the San Ysidro neighborhood at the San Ysidro Port of Entry.[86] A second, primarily commercial border crossing operates in the Otay Mesa area; it is the largest commercial crossing on the California-Baja California border and handles the third highest volume of trucks and dollar value of trade among all United States-Mexico land crossings.[87]

One of the Port of San Diego's two cargo facilities is located in Downtown San Diego at the Tenth Avenue Marine Terminal. This terminal has facilities for containers, bulk cargo, and refrigerated and frozen storage, so that it can handle the import and export of perishables (including 33 million bananas every month) as well as fertilizer, cement, forest products, and other commodities.[88] In 2009 the Port of San Diego handled 1,137,054 short tons of total trade; foreign trade accounted for 956,637 short tons while domestic trade amounted to 180,417 short tons.[89]

Manufacturing and research

In 2010, former Governor Schwarzenegger's Office of Economic Development designated San Diego as an iHub Innovation Center for collaboration potentially between wireless and life sciences, citing the area's wireless business, pharmaceutical research and start-ups for medical devices and diagnostics.[90]

Qualcomm corporate headquarters

San Diego hosts several major producers of wireless cellular technology. Qualcomm was founded and is headquartered in San Diego, and still is the largest private-sector technology employer (excluding hospitals) in San Diego County.[91] Other wireless industry manufacturers headquartered here include LG Electronics,[92] Kyocera International.,[93] and Novatel Wireless.[94] According to the *San Diego Business Journal*, the largest software company in San Diego is security software company Websense Inc.[95] San Diego also has the U.S. headquarters for the Slovakian security company ESET.[96]

The presence of the University of California, San Diego and other research institutions has helped to fuel biotechnology growth.[97] In June 2004, San Diego was ranked the top biotech cluster in the United States by the Milken Institute.[98] There are more than 400 biotechnology companies in the area.[99] In particular, the La Jolla and nearby Sorrento Valley areas are home to offices and research facilities for numerous biotechnology companies.[100] Major biotechnology companies like Neurocrine Biosciences and Nventa Biopharmaceuticals are headquartered in San Diego, while many biotech and pharmaceutical companies, such as BD Biosciences, Biogen Idec, Integrated DNA Technologies, Merck, Pfizer, Élan, Celgene, and Vertex, have offices or research facilities in San Diego. There are also several non-profit biotech and health care institutes, such as the Salk Institute for Biological Studies, the Scripps Research Institute, the West Wireless Health Institute and the Sanford-Burnham Institute. San Diego is also home to more than 140 contract research organizations (CROs) that provide a variety of contract services for pharmaceutical and biotechnology companies.[101]

Real estate

Skyline view of the Village of La JollaVillage of La Jolla in San Diego

Prior to 2006, San Diego experienced a dramatic growth of real estate prices, to the extent that the situation was sometimes described as a "housing affordability crisis". Median house prices more than tripled between 1998 and 2007. According to the California Association of Realtors, in May 2007, a median house in San Diego cost $612,370.[102] Growth of real estate prices has not been accompanied by comparable growth of household incomes: Housing Affordability Index (percentage of households that can afford to buy a median-priced house) fell below 20 percent in the early 2000s. The San Diego metropolitan area had the second worst median multiple (ratio of median house price to median household income) of all metropolitan areas in the United States. As a consequence, San Diego had experienced negative net migration since 2004, with significant numbers of people moving to Baja California and Riverside County, with many residents commuting daily from Tijuana, Temecula, and Murrieta, to their jobs in San Diego. Others are leaving the state altogether and moving to more affordable regions.[103]

San Diego home prices peaked in 2005 then declined as part of a nationwide trend. As of December 2010, home prices were 60 percent higher than in 2000, but down 36 percent from the peak in 2005.[104] The median home price declined by more than $200,000 between 2005 and 2010, and sales dropped by 50 percent.[105]

Top employers

According to the City's 2009 Comprehensive Annual Financial Report,[106] the top employers in the city are:

The United States Navy is San Diego's largest employer.

Employer	Number of employees
United States Navy	55,300
San Diego Unified School District	21,959
University of California, San Diego	19,435
San Diego County	17,900
Sharp HealthCare	14,724
City of San Diego	10,799
Kaiser Permanente	7,220
University of San Diego	6,086
Qualcomm	6,000
UCSD Medical Center	5,300

Education

Primary and secondary schools

Public schools in San Diego are operated by independent school districts. The majority of the public schools in the city are served by the San Diego Unified School District, also the second largest school district in California, which includes 11 K-8 schools, 107 elementary schools, 24 middle schools, 13 atypical and alternative schools, 28 high schools, and 45 charter schools.[107] Several adjacent school districts which are headquartered outside the city limits serve some schools within the city; these include the Poway Unified School District, Del Mar Union School District, San Dieguito Union High School District and Sweetwater Union High School District. In addition, there are a number of private schools in the city.

Colleges and universities

According to education rankings released by the U.S. Census Bureau, 40.4 percent of San Diegans ages 25 and older hold bachelor's degrees. The census ranks the city as the ninth most educated city in the United States based on these figures.[108]

San Diego State University's Hepner Hall

Public colleges and universities in the city include San Diego State University (SDSU), University of California, San Diego (UCSD), and the San Diego Community College District, which includes San Diego City College, San Diego Mesa College, and San Diego Miramar College. Private colleges and universities in the city include University of San Diego (USD), Point Loma Nazarene University (PLNU), Alliant International University (AIU), National University, California International Business University (CIBU), San Diego Christian College, John Paul the Great Catholic University, California College San Diego, Coleman University, University of Redlands School of Business, Design Institute of San Diego (DISD), Fashion Institute of Design & Merchandising's San Diego campus, NewSchool of Architecture and Design, Pacific Oaks College San Diego Campus, Chapman University's San Diego Campus, The Art Institute of California-San Diego, Southern States University (SSU), UEI College, and Woodbury University School of Architecture's satellite campus.

There is one medical school in the city, the UCSD School of Medicine. There are three ABA accredited law schools in the city, which include California Western School of Law, Thomas Jefferson School of Law, and University of San Diego School of Law. There is also one unaccredited law school, Western Sierra Law School.

Libraries

University of California, San Diego's Geisel Library, named for Theodor Seuss Geisel ("Dr. Seuss")

The city-run San Diego Public Library system is headquartered downtown and has 34 branches throughout the city.[109] The libraries have had reduced operating hours since 2003 due to the city's financial problems. In 2006 the city increased spending on libraries by $2.1 million.[110] In the new 2011 budget, the mayor proposed further reducing library hours to 18 hours per week.[111]

In addition to the municipal public library system, there are nearly two dozen libraries open to the public which are run by other governmental agencies and by schools, colleges, and universities.[112] Noteworthy among them are the Malcolm A. Love Library at San Diego State University and the Geisel Library at the University of California, San Diego.

Culture

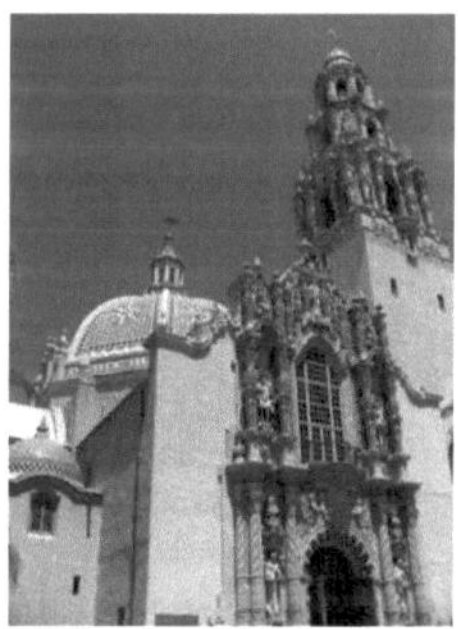

The Museum of Man

Many popular museums, such as the San Diego Museum of Art, the San Diego Natural History Museum, the San Diego Museum of Man, and the Museum of Photographic Arts are located in Balboa Park. The Museum of Contemporary Art San Diego (MCASD) is located in La Jolla and has a branch located at the Santa Fe Depot downtown. The Columbia district downtown is home to historic ship exhibits belonging to the San Diego Maritime Museum, headlined by the Star of India, as well as the unrelated San Diego Aircraft Carrier Museum featuring the USS Midway aircraft carrier.

The San Diego Symphony at Symphony Towers performs on a regular basis and is directed by Jahja Ling. The San Diego Opera at Civic Center Plaza, directed by Ian Campbell, was ranked by Opera America as one of the top 10 opera companies in the United States. Old Globe Theatre at Balboa Park produces about 15 plays and musicals annually. The La Jolla Playhouse at UCSD is directed by Christopher Ashley. Both the Old Globe Theatre and the La Jolla Playhouse have produced the world premieres of plays and musicals that have gone on to win Tony Awards[113] or nominations[114] on Broadway. The Joan B. Kroc Theatre at Kroc Center's Performing Arts Center is a 600-seat state-of-the-art theatre that hosts music, dance, and theatre performances. The San Diego Repertory Theatre at the Lyceum Theatres in Horton Plaza produces a variety of plays and musicals. Other professional theatrical production companies include the Lyric Opera San Diego and the Starlight Musical Theatre. Hundreds of movies and a dozen TV shows have been filmed in San Diego, a tradition going back as far as 1898.[115] [116]

Sports

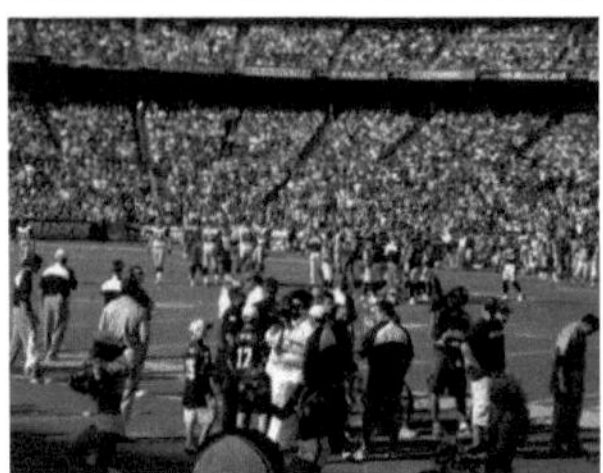

Qualcomm Stadium (formerly named "Jack Murphy Stadium" after a sports writer) hosts a San Diego Chargers game with the St. Louis Rams.

The National Football League's San Diego Chargers play in Qualcomm Stadium. Three NFL Super Bowl championships have been held there. Major League Baseball's San Diego Padres play in Petco Park. Parts of the World Baseball Classic were played there in 2006 and 2009.

NCAA Division I San Diego State Aztecs men's and women's basketball games are played at Viejas Arena at Aztec Bowl on the campus of San Diego State University. College football and soccer, basketball and volleyball are played at the Torero Stadium and the Jenny Craig Pavilion at USD.

The San Diego State Aztecs (MWC) and the University of San Diego Toreros (WCC) are NCAA Division I teams. The UCSD Tritons (CCAA) are members of NCAA Division II while the Point Loma Nazarene Sea Lions and San Diego Christian College (GSAC) are members of the NAIA.

Petco Park

Qualcomm stadium also houses the NCAA Division I San Diego State Aztecs, as well as local high school football championships, international soccer games, and supercross events. Two of college football's annual bowl games are also held there: the Holiday Bowl and the Poinsettia Bowl. Soccer, American football, and track and field are played in Balboa Stadium, the city's first stadium, constructed in 1914.

Rugby union is a developing sport in the city. The USA Sevens, a major rugby event, was held there from 2007 through 2009. San Diego is one of only 16 cities in the United States included in the Rugby Super League[117] represented by Old Mission Beach Athletic Club RFC, the home club of USA Rugby's Captain Todd Clever who plays rugby professionally in Japan's Top League with Suntory Sungoliath.[118] San Diego will participate in the Western American National Rugby League which starts in 2011.[119]

The San Diego Surf of the American Basketball Association is located in the city. The annual Farmers Insurance Open golf tournament (formerly the Buick Invitational) on the PGA Tour occurs at the municipally owned Torrey Pines Golf Course. This course was also the site of the 2008 U.S. Open Golf Championship. The San Diego Yacht Club hosted the America's Cup yacht races three times during the period 1988 to 1995. The amateur beach sport Over-the-line was invented in San Diego,[120] and the annual world Over-the-line championships are held at Mission Bay every year.[121]

Media

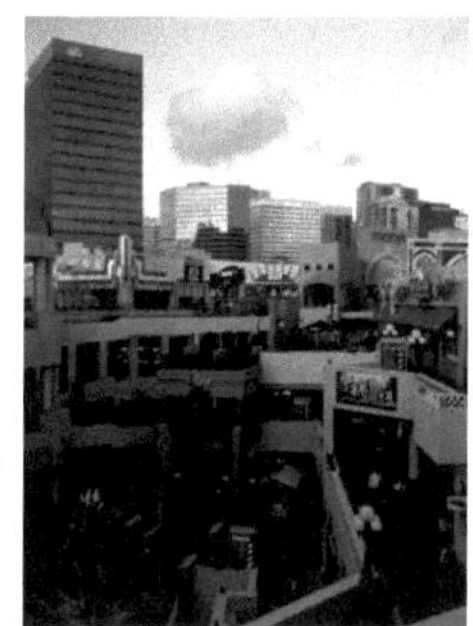
NBC San Diego (left) is outside Horton Plaza on Broadway downtown.

The following are published within the city: the daily newspaper, *The San Diego Union-Tribune* and its online portal, *signonsandiego.com*,[122] and the alternative newsweeklies, the *San Diego CityBeat* and *San Diego Reader*. Another newspaper is the *North County Times*, which is distributed in San Diego's North County area. *Voice of San Diego* is a non-profit online-only news outlet covering government, politics, education, neighborhoods, and the arts. The *San Diego Daily Transcript* is a business-oriented daily newspaper.

San Diego led U.S. local markets with 69.6 percent broadband penetration in 2004 according to Nielsen//NetRatings.[123]

San Diego's first television station was KFMB, which began broadcasting on May 16, 1949.[124] Since the Federal Communications Commission (FCC) licensed seven television stations in Los Angeles, two VHF channels were available for San Diego because of its relative proximity to the larger city. In 1952, however, the FCC began licensing UHF channels, making it possible for cities such as San Diego to acquire more stations. Stations based in Mexico (with ITU prefixes of XE and XH) also serve the San Diego market. Television stations today include XHTJB 3 (Once TV), XETV 6 (CW), KFMB 8 (CBS), KGTV 10 (ABC), XEWT 12 (Televisa Regional), KPBS 15 (PBS), KBNT 17 (Univision), XHTIT 21 (Azteca 7), XHJK 27 (Azteca 13), KSDX-LP 29 (Spanish Independent), XHAS 33 (Telemundo), K35DG 35 (UCSD-TV), KDTF 36 (Telefutura), KNSD 39 (NBC), KZSD-LP 41 (Azteca America), KSEX-CA 43 (Infomercials), XHBJ 45 (Canal 5), XHDTV 49 (MNTV), KUSI 51 (Independent), XHUAA 57 (Canal de las Estrellas),and KSWB-TV 69 (Fox). San Diego has an 80.6 percent cable penetration rate.[125]

The radio stations in San Diego include nationwide broadcaster, Clear Channel Communications; CBS Radio, Midwest Television, Lincoln Financial Media, Finest City Broadcasting, and many other smaller stations and networks. Stations include: KOGO AM 600, KFMB AM 760, KCEO AM 1000, KCBQ AM 1170, K-Praise, KLSD AM 1360 *Air America*, KFSD 1450 AM, KPBS-FM 89.5, Channel 933, Star 94.1, FM 94/9, New Country 95.7, Q96 96.1, KyXy 96.5, Free Radio San Diego (AKA Pirate Radio San Diego) 96.9FM FRSD, KSON 97.3/92.1, KIFM 98.1, Jack-FM 100.7, 101.5 KGB-FM, KPRI 102.1, Rock 105.3, and another *Pirate Radio* station at 106.9FM, as well as a number of local Spanish language radio stations.

Government

Local government

San Diego City Council chambers

The city is governed by a mayor and an 8-member city council. In 2006, the city's form of government changed from a "city manager system" to a "strong mayor system". The change was brought about by a citywide vote in 2004. The mayor is in effect the chief executive officer of the city, while the council is the legislative body.[126]

The members of the city council are each elected from single member districts within the city. The mayor and city attorney are elected directly by the voters of the entire city. The mayor, city attorney, and council members are elected to four-year terms, with a two-term limit.[127] Elections are held on a non-partisan basis per California state law; nevertheless, most officeholders do identify themselves as either Democrats or Republicans.

Mayor Jerry Sanders

Although San Diego has a Republican mayor,[128] in 2007, registered Democrats outnumbered Republicans by about 7 to 6 in the city, and Democrats currently hold a 5–3 majority in the City Council.[129]

San Diego is part of San Diego County which is governed by a 5-member Board of Supervisors. The City of San Diego includes all or part of the 1st, 2nd, 3rd and 4th supervisorial[130] districts, represented by Republican County Supervisors Greg Cox, Dianne Jacob, Pam Slater-Price and Ron Roberts.[131]

Areas of the city immediately adjacent to San Diego Bay ("tidelands") are administered by the Port of San Diego, a quasi-governmental agency which owns all the property in the tidelands and is responsible for its land use planning, policing, and similar functions. San Diego is a member of the regional planning agency San Diego Association of Governments (SANDAG). Public schools within the city are managed and funded by independent school districts (see above).

State and federal

In the state legislature San Diego is located in the 36th, 38th, 39th, and 40th Senate District, represented by Republicans Joel Anderson and Mark Wyland, and Democrats Christine Kehoe and Juan Vargas, and in the 74th, 75th, 76th, 77th, 78th, and 79th Assembly District, represented by Republicans Martin Garrick and Nathan Fletcher, Democrat Toni Atkins, Republican Brian Jones and Democrats Marty Block and Ben Hueso, respectively.

Federally, San Diego is located in California's 49th, 50th, 51st, 52nd, and 53rd congressional districts, which have Cook PVIs of R +10, R +5, D +7, R +9, and D +12 respectively[132] and are represented by Republicans Darrell Issa and Brian Bilbray, Democrat Bob Filner, Republican Duncan D. Hunter, and Democrat Susan Davis, respectively.

Major scandals

Then-mayor Roger Hedgecock was forced to resign his post in 1985, after he was found guilty of one count of conspiracy and twelve counts of perjury, related to the alleged failure to report all campaign contributions.[133] [134] After a series of appeals, the twelve perjury counts were dismissed in 1990 based on claims of juror misconduct; the remaining conspiracy count was reduced to a misdemeanor and then dismissed.[135]

A 2002 scheme to underfund pensions for city employees led to the San Diego pension scandal. This resulted in the resignation of newly re-elected Mayor Dick Murphy[136] and the criminal indictment of six pension board members.[137] Those charges were finally dismissed by a federal judge in 2010.[138]

On November 28, 2005, U.S. Congressman Randy "Duke" Cunningham resigned after being convicted on federal bribery charges. He had represented California's 50th congressional district, which includes much of the northern portion of the city of San Diego. In 2006, Cunningham was sentenced to a 100-month prison sentence.[139]

In 2005 two city council members, Ralph Inzunza and Deputy Mayor Michael Zucchet — who briefly took over as Acting Mayor when Murphy resigned — were convicted of extortion, wire fraud, and conspiracy to commit wire fraud for taking campaign contributions from a strip club owner and his associates, allegedly in exchange for trying to repeal the city's "no touch" laws at strip clubs.[140] Both subsequently resigned. In 2009, a judge acquitted Zucchet on seven out of the nine counts against him, and granted his petition for a new trial on the other two charges;[141] the remaining charges were eventually dropped.[142]

Infrastructure

Utilities

Water is supplied to residents by the Water Department of the City of San Diego. The city receives its water from the Metropolitan Water District of Southern California.

Gas and electric utilities are provided by San Diego Gas & Electric, a division of Sempra Energy.

Transportation

I-5 looking south toward downtown San Diego

With the automobile being the primary means of transportation for over 80 percent of its residents, San Diego is served by a network of freeways and highways. This includes Interstate 5, which runs south to Tijuana and runs north to Los Angeles; Interstate 8, which runs east to Imperial County and the Arizona Sun Corridor; Interstate 15, which runs northeast through the Inland Empire to Las Vegas; and Interstate 805, which splits from I-5 near the Mexican border and rejoins I-5 at Sorrento Valley.

Major state highways include SR 94, which connects downtown with I-805, I-15 and East County; SR 163, which connects downtown with the northeast part of the city, intersects I-805 and merges with I-15 at Miramar; SR 52, which connects La Jolla with East County through Santee and SR 125; SR 56, which connects I-5 with I-15 through Carmel Valley and Rancho Peñasquitos; SR 75, which spans San Diego Bay as the San Diego-Coronado Bridge, and also passes through South San Diego as Palm Avenue; and SR 905, which connects I-5 and I-805 to the Otay Mesa Port of Entry.

The stretch of SR 163 that passes through Balboa Park is San Diego's oldest freeway, and has been called one of America's most beautiful parkways.[143]

San Diego's roadway system provides an extensive network of routes for travel by bicycle. The dry and mild climate of San Diego makes cycling a convenient and pleasant year-round option. At the same time, the city's hilly, canyon-like terrain and significantly long average trip distances—brought about by strict low-density zoning laws—somewhat restrict cycling for utilitarian purposes. Older and denser neighborhoods around the downtown tend to be utility cycling oriented. This is partly because of the grid street patterns now absent in newer developments farther from the urban core, where suburban style arterial roads are much more common. As a result, a vast majority of cycling-related activities are recreational. Testament to San Diego's cycling efforts, in 2006, San Diego was rated as the best city for cycling for U.S. cities with a population over 1 million.[144]

San Diego is served by the trolley, bus, Sprinter, Coaster, and Amtrak. The trolley primarily serves downtown and surrounding urban communities, Mission Valley, east county, and coastal south bay. A planned Mid-Coast line will operate from Old Town to University City along the 5 Freeway. The Amtrak and Coaster trains currently run along the coastline and connect San Diego with Los Angeles, Orange County, Riverside, San Bernardino, and Ventura via Metrolink. There are two Amtrak stations in San Diego, in Old Town and Downtown. San Diego transit information about public transportation and commuting is available on the Web and by dialing "511" from any phone in the area.[145] [146]

View of Coronado and San Diego from the air

The city's primary commercial airport is the San Diego International Airport (SAN), also known as Lindbergh Field. It is the busiest single-runway airport in the United States,[147] that served over 17 million passengers in 2005, and is dealing with an increasingly larger number every year.[147] It is located on San Diego Bay three miles (4.8 km) from downtown. San Diego International Airport maintains scheduled flights to the rest of the United States including Hawaii, as well as to Mexico and Canada. It is operated by an independent agency, the San Diego Regional Airport Authority. In addition, the city itself operates two general-aviation airports, Montgomery Field (MYF) and Brown Field (SDM).[148]

Several regional transportation projects have been undertaken in recent years to deal with congestion in San Diego. Notable efforts are on San Diego freeways, San Diego Airport, and the cruise terminal of the port. Freeway projects include expansion of Interstates 5 and 805 around "The Merge," a rush-hour spot where the two freeways meet. Also, an expansion of Interstate 15 through the North County is underway with the addition of high-occupancy-vehicle (HOV) "managed lanes". There is a tollway (The South Bay Expressway) connecting SR 54 and Otay Mesa, near the Mexican border. According to a 2007 assessment, 37 percent of streets in San Diego were in acceptable driving condition. The proposed budget fell $84.6 million short of bringing the city's streets to an acceptable level.[149] Port expansions included a second cruise terminal on Broadway Pier which opened in 2010. Airport projects include expansion of Terminal 2, currently under construction and slated for completion in 2013.[150]

Walkability

A 2011 study by Walk Score ranked San Diego the eighteenth most walkable of fifty largest cities in the United States.[151]

Sister cities

San Diego has 16 sister cities, as designated by Sister Cities International:[152] [153]

- Alcalá de Henares, Spain
- Campinas, Brazil
- Cavite, Philippines
- Edinburgh, Scotland, United Kingdom
- Jalalabad, Afghanistan
- Jeonju, South Korea
- León, Mexico

- Perth, Australia
- Quanzhou, People's Republic of China[153]
- Taichung City, Taiwan (Republic of China)
- Tema, Ghana
- Tijuana, Mexico
- Vladivostok, Russia
- Warsaw, Poland[154]
- Yantai, People's Republic of China
- Yokohama, Japan[155]

See also

- 1858 San Diego hurricane
- List of notable San Diegans
- List of California public officials charged with crimes, San Diego
- Category:Visitor attractions in San Diego County, California
- Category: Museums in San Diego County

References

[1] U.S. Census (http://www.census.gov/geo/www/gazetteer/files/Gaz_places_national.txt)

[2] http://www.sandiego.gov/

[3] McGrew, Clarence Alan (1922). *City of San Diego and San Diego County: the birthplace of California* (http://books.google.com/books/about/City_of_San_Diego_and_San_Diego_County.html?id=nc8KiryvkdYC). American Historical Society. . Retrieved July 23, 2011.

[4] "CITY OF SAN DIEGO 2010 REDISTRICTING COMMISSION OVERVIEW OF 2010 U.S. CENSUS DATA" (http://www.sandiego.gov/redistricting/pdf/overviewuscensusdata.pdf). Redistricting Commission - City of San Diego. . Retrieved 24 August 2011.

[5] "Kumeyaay indians" (http://www.kumeyaay.info/kumeyaay_indians.html). kumeyaay.info. . Retrieved July 1, 2010.

[6] "San Diego Historical Society" (https://www.sandiegohistory.org/bio/cabrillo/cabrillo.htm). Sandiegohistory.org. . Retrieved March 12, 2011.

[7] "Journal of San Diego History, October 1967" (https://www.sandiegohistory.org/journal/67october/began.htm). Sandiegohistory.org. . Retrieved March 12, 2011.

[8] "San Diego Historical Society:Timeline of San Diego history" (http://www.sandiegohistory.org/timeline/timeline1.htm). Sandiegohistory.org. . Retrieved May 4, 2011.

[9] "Keyfacts" (http://www.missionscalifornia.com/keyfacts/san-diego-de-alcala.html). missionscalifornia.com. . Retrieved July 1, 2010.

[10] "Mission San Diego" (http://www.missionsandiego.com/). Mission San Diego. . Retrieved July 1, 2010.

[11] "National Park Service, National Historicl Landmarks Program: San Diego Presidio" (http://tps.cr.nps.gov/nhl/detail.cfm?ResourceID=130&resourceType=Site). Tps.cr.nps.gov. October 10, 1960. . Retrieved May 4, 2011.

[12] "City of San Diego website" (http://www.sandiego.gov/city-clerk/officialdocs/legisdocs/charter.shtml). Sandiego.gov. . Retrieved July 1, 2010.

[13] *Engstrand, Iris Wilson, California's Cornerstone, Sunbelt Publications, Inc., 2005, p. 80* (http://books.google.com/?id=RhCQUf1XQ84C&pg=PA87&lpg=PA87&dq=san+diego+"new+town"+horton&q=san diego "new town" horton). May 30, 2005. ISBN 9780932653727. . Retrieved July 1, 2010.

[14] Steele, Jeanette (May 1, 2005). "Balboa Park future is full of repair jobs" (http://pqasb.pqarchiver.com/sandiego-sub/access/831435001.html?dids=831435001:831435001&FMT=FT&FMTS=ABS:FT&type=current&date=May+1,+2005&author=Jeanette+Steele&pub=The+San+Diego+Union+-+Tribune&edition=&startpage=B.1&desc=Balboa+Park+future+is+full+of+repair+jobs+|+Projects+may+mean+detours+for+visitors). *The San Diego Union*. . Retrieved July 1, 2010.

[15] Marjorie Betts Shaw,. "The San Diego Zoological Garden: A Foundation to Build on" (http://www.sandiegohistory.org/journal/78summer/zoo.htm). *Journal of San Diego History*. Sandiegohistory.org. . Retrieved May 4, 2011.

[16] University of San Diego: Military Bases in San Diego (http://history.sandiego.edu/gen/local/kearny/page00d.html)

[17] Gerald A. Shepherd,. "When the Lone Eagle returned to San Diego" (http://www.sandiegohistory.org/journal/94winter/eagle.htm). *Journal of San Diego History*. Sandiegohistory.org. . Retrieved May 4, 2011.

[18] Moffatt, Riley. *Population History of Western U.S. Cities & Towns, 1850–1990*. Lanham: Scarecrow, 1996, 54.

[19] "Milken Institute" (http://www.milkeninstitute.org/publications/publications.taf?function=detail&ID=312&cat=resrep). Milken Institute. . Retrieved July 1, 2010.

[20] Erie, Steven P.; Kogan, Vladimir; MacKenzi, Scott A. (January 27, 2010). "Redevelopment, San Diego Style: The Limits of Public–Private Partnerships" (http://uar.sagepub.com/content/45/5/644). *Urban Affairs Review* **45** (5): 644– 678. doi:10.1177/1078087409359760. . Retrieved November 4, 2010.
[21] Schad, Jerry. *Afoot and Afield in San Diego* (http://books.google.com/books?id=AqZUHkIaSXYC&pg=PT129&dq=san+diego+canyons+neighborhoods#v=onepage&q=san diego canyons neighborhoods&f=false). Wilderness Press, Berkeley, Calif.. p. 111. . Retrieved May 4, 2011.
[22] "City of San Diego Community Planning Areas" (http://www.sandiego.gov/planning/community/profiles/index.shtml). Sandiego.gov. . Retrieved May 4, 2011.
[23] "City of San Diego" (http://www.sandiego.gov/neighborhoodmaps/). Sandiego.gov. . Retrieved July 1, 2010.
[24] *Aitken, Stuart, and Prosser, Rudy, ',Residents' Spatial Knowledge of Neighborhood Continuity and Form',, Geographical Analysis, September 3, 2010.* September 3, 2010. doi:10.1111/j.1538-4632.1990.tb00213.x.
[25] Roger Showley (April 18, 2010). "City, SANDAG win planning awards" (http://www.signonsandiego.com/news/2010/apr/18/city-sandag-win-planning-awards/). *San Diego Union-Tribune*. . Retrieved May 4, 2011.
[26] Engstrand, Iris Wilson, *California's Cornerstone* (http://books.google.com/books?id=RhCQUf1XQ84C&pg=PA87&lpg=PA87&dq=san+diego+"new+town"+horton#v=onepage&q=san diego "new town" horton&f=false). Sunbelt Publications, Inc.. 2005. p. 80. Engstrand, Iris Wilson,. Retrieved May 4, 2011.
[27] "San Diego Timeline Diagram" (http://skyscraperpage.com/diagrams/?cityID=120&searchname=timeline). Skyscraper Source Media. . Retrieved May 31, 2011.
[28] "One America Plaza" (http://www.emporis.com/en/wm/bu/?id=1americaplaza-sandiego-ca-usa). Emporis.com. . Retrieved May 16, 2009.
[29] "Airport Land Use Compatibility Plan for San Diego International Airport" (http://www.san.org/documents/aluc/SDIA_ALUCP.pdf) (PDF). San Diego County Regional Airport Authority. October 4, 2004. pp. 51–52. . Retrieved May 16, 2009.
[30] Bergman, Heather (June 27, 2005). "San Diego's skyline grows up: residential towers filling some of the missing 'tools' as office projects are nearing completion" (http://www.thefreelibrary.com/San+Diego's+skyline+grows+up:+residential+towers+filling+some+of+the...-a0134576016). *San Diego Business Journal* (TheFreeLibrary.com). . Retrieved May 16, 2009.
[31] Geiger, Peter (October 5, 2006). "The 10 Best Weather Cities" (http://www.farmersalmanac.com/blog/2006/10/05/the-10-best-weather-cities/). *Farmer's Almanac* (Almanac Publishing). . Retrieved April 19, 2011.
[32] Kellogg, Becky and Erdman, Jonathan (September 2010). "America's Best Climates" (http://wwworigin.weather.com/outlook/weather-news/news/articles/americas-best-climates-poll_2010-10-19?page=3). The Weather Channel. . Retrieved April 19, 2011.
[33] M. Kottek; J. Grieser, C. Beck, B. Rudolf, and F. Rubel (2006). "World Map of the Köppen-Geiger climate classification updated" (http://koeppen-geiger.vu-wien.ac.at/pics/kottek_et_al_2006.gif). *Meteorol. Z.* **15**: 259–263. doi:10.1127/0941-2948/2006/0130. . Retrieved April 22, 2009.
[34] "UCSD" (http://meteora.ucsd.edu/cap/gloom.html). Meteora.ucsd.edu. May 14, 2010. . Retrieved July 1, 2010.
[35] "Monthly Averages for San Diego, CA" (http://www.weather.com/outlook/travel/businesstraveler/wxclimatology/monthly/USca0982). The Weather Channel. . Retrieved April 22, 2009.
[36] "Monthly Averages for El Cajon, CA" (http://www.weather.com/outlook/travel/businesstraveler/wxclimatology/monthly/92020). The Weather Channel. . Retrieved April 22, 2009.
[37] Lee, Mike (June 18, 2011). "Is global warming changing California Current?" (http://www.signonsandiego.com/news/2011/jun/18/taking-stock-california-current/). *U-T (San Diego Union Tribune)*. . Retrieved June 20, 2011.
[38] *"San Diego's average rainfall set to lower level"* (http://www.signonsandiego.com/news/2011/mar/16/san-diegos-average-rainfall-set-lower-level/). *San Diego Union-Tribune*. March 16, 2011. . Retrieved April 12, 2011.
[39] Rowe, Peter (December 13, 2007). "The day it snowed in San Diego" (http://legacy.signonsandiego.com/news/metro/20071213-9999-1n13snowday.html). *San Diego Union Tribune*. . Retrieved May 4, 2011.
[40] "National Oceanic and Atmospheric Agency: San Diego climate by month" (http://www.wrh.noaa.gov/sgx/climate/san-san-month.htm). Wrh.noaa.gov. . Retrieved May 4, 2011.
[41] Conner, Glen. History of weather observations San Diego, California 1849—1948 (http://mrcc.sws.uiuc.edu/FORTS/histories/CA_San_Diego_Conner.pdf). Climate Database Modernization Program, NOAA's National Climatic Data Center. pp. 7–8.
[42] "Climatography of the United States No. 20: San Diego Lindbergh AP, CA (1971–2000)" (http://cdo.ncdc.noaa.gov/climatenormals/clim20/ca/047740.pdf) (PDF). National Oceanic and Atmospheric Administration. 2004. . Retrieved 2010-05-31.
[43] "Climatological Normals of San Diego" (http://www.hko.gov.hk/wxinfo/climat/world/eng/n_america/us/san_diego_e.htm). Hong Kong Observatory. . Retrieved 2010-05-12.
[44] Wells, Michael L.; John F. O'Leary, Janet Franklin, Joel Michaelsen, and David E. McKinsey (November 2, 2004). "Variations in a regional fire regime related to vegetation type in San Diego County, California (USA)" (http://www.springerlink.com/content/xx00155q65147l45/). *Landscape Ecology* (San Diego, CA 92182-4493, USA: Springer Netherlands) **19** (2): 139–152. doi:10.1023/B:LAND.0000021713.81489.a7. 1572-9761. . Retrieved April 22, 2009.
[45] Strömberg, Nicklas; Michael Hogan (November 29, 2008). "Torrey Pine: Pinus torreyana" (http://globaltwitcher.auderis.se/artspec.asp?thingid=62498). GlobalTwitcher. . Retrieved April 22, 2009.
[46] "Tecolote Canyon Natural Park & Nature Center" (http://www.sandiego.gov/park-and-recreation/parks/teclte.shtml). The City of San Diego. . Retrieved April 22, 2009.

[47] "Marian Bear Memorial Park" (http://www.sandiego.gov/park-and-recreation/parks/marbear2.shtml). The City of San Diego. . Retrieved April 22, 2009.

[48] Lee, Mike (March 28, 2007). "White House seeks limits to species act" (http://www.signonsandiego.com/news/politics/20070328-9999-1n28esa.html). *San Diego Union-Tribune*. . Retrieved April 22, 2009.

[49] "San Diego County Bird Atlas Project" (http://www.sdnhm.org/research/birdatlas_draft/index.html). San Diego Natural History Museum. .

[50] "Corpus Christi Recognized as Birdiest City" (http://www.corpuschristidaily.com/article_detail_new.cfm?id=1353). *Corpus Christi Daily*. December 2004. . Retrieved April 13, 2011.

[51] "Corpus Christi remains 'birdiest city in America'" (http://www.scribd.com/doc/36229005/Corpus-Christi-Remains-'Birdiest-City-in-America). Corpus Christi Convention and Visitors Bureau. June 25, 2008. . Retrieved April 13, 2011.

[52] Goldstein, Bruce Evan (September 2007). "The Futility of Reason: Incommensurable Differences Between Sustainability Narratives in the Aftermath of the 2003 San Diego Cedar Fire" (http://www.informaworld.com/smpp/content~content=a787467532~db=all). *Journal of Environmental Policy & Planning* (Blacksburg, USA: School of Public and International Affairs, Virginia Tech) **9** (3 & 4): 227–244. doi:10.1080/15239080701622766. . Retrieved April 22, 2009.

[53] "CalFire website" (http://www.fire.ca.gov/cdf/incidents/Cedar Fire_120/incident_info.html). Fire.ca.gov. . Retrieved July 1, 2010.

[54] Viswanathan, S.; L. Eria, N. Diunugala, J. Johnson, C. McClean (January 2006). "An Analysis of Effects of San Diego Wildfire on Ambient Air Quality" (http://md1.csa.com/partners/viewrecord.php?requester=gs&collection=ENV&recid=6707765&q=wildfire+"san+diego+"&uid=&setcookie=yes). *Journal of the Air & Waste Management Association* **56** (1). . Retrieved December 15, 2008.

[55] Manolatos, Tony (October 22, 2007). "Wildfires seen as eclipsing the Cedar fire of 2003" (http://www.signonsandiego.com/news/metro/20071022-0922-bn22fire3.html). *San Diego Union Tribune*. Signonsandiego.com. . Retrieved July 1, 2010.

[56] "Annual Estimates of the Resident Population for Incorporated Places Over 100,000, Ranked by July 1, 2009 Population: April 1, 2000 to July 1, 2009" (http://www.census.gov/popest/cities/tables/SUB-EST2009-01.csv) (CSV). United States Census Bureau, Population Division. July 1, 2009. . Retrieved June 28, 2010.

[57] Census: 1,307,402 Live in San Diego (March 8, 2011). "Voice of San Diego, March 8, 2011" (http://www.voiceofsandiego.org/data-drive/article_0b4c5ece-49cd-11e0-be00-001cc4c002e0.html). Voiceofsandiego.org. . Retrieved May 4, 2011.

[58] "San Diego (city) QuickFacts from the US Census Bureau" (http://quickfacts.census.gov/qfd/states/06/0666000.html). US Census Bureau. . Retrieved February 14, 2010.

[59] "2010 Census P.L. 94-171 Summary File Data" (http://www2.census.gov/census_2010/01-Redistricting_File--PL_94-171/California/). United States Census Bureau. .

[60] "San Diego, CA Census Profile" (http://www.usatoday.com/news/nation/census/profile/ca). *USA Today*. March 8, 2011. . Retrieved March 12, 2011.

[61] "Population and Housing Estimates" (http://profilewarehouse.sandag.org/profiles/est/city14est.pdf) (PDF). SANDAG: Profile Warehouse (http://profilewarehouse.sandag.org/). 2008. . Retrieved April 22, 2009.

[62] "Census Quick Facts, City of San Diego" (http://quickfacts.census.gov/qfd/states/06/0666000.html). Quickfacts.census.gov. . Retrieved July 1, 2010.

[63] "City of San Diego Economic Development Department" (http://www.sandiego.gov/economic-development/glance/sdfacts.shtml). Sandiego.gov. . Retrieved July 1, 2010.

[64] "SANDAG document" (http://docs.google.com/gview?a=v&q=cache:jmr6Ynim0y4J:profilewarehouse.sandag.org/profiles/est/city14est.pdf+city+san+diego+population+age&hl=en&gl=us). Google. . Retrieved July 1, 2010.

[65] "San Diego city, California" (http://factfinder.census.gov/servlet/SAFFFacts?_event=&geo_id=16000US0666000&_geoContext=01000US). United States Census Bureau. 2000. . Retrieved April 22, 2009.

[66] Clemence, Sara (October 28, 2005). "Richest Cities In The U.S." (http://www.forbes.com/2005/10/27/richest-cities-US-cx_sc_1028home_ls.html). *Forbes*. . Retrieved April 22, 2009.

[67] "Best Places to Live 2006" (http://money.cnn.com/magazines/moneymag/bplive/2006/snapshots/PL0666000.html). *Money*. 2006. . Retrieved November 29, 2009.

[68] Levy, Francesca (September 11, 2010). "America's Safest Cities" (http://www.forbes.com/2010/10/11/safest-cities-america-crime-accidents-lifestyle-real-estate-danger.html). *Forbes*. . Retrieved February 20, 2011.

[69] "SDPD Historical Crime Actuals 1950–2006" (http://www.sandiego.gov/police/pdf/UCR50to2006.pdf) (PDF). San Diego Police Department. April 14, 2006. . Retrieved April 22, 2009.

[70] "SDPD Historical Crime Rates Per 1,000 Population 1950–2006" (http://www.sandiego.gov/police/pdf/UCRrates50to2006.pdf) (PDF). San Diego Police Department. April 14, 2006. .

[71] Manolatos, Tony; Kristina Davis (April 14, 2006). "County crows at glowing crime report" (http://www.signonsandiego.com/uniontrib/20060414/news_7m14stats.html). *San Diego Union-Tribune*. . Retrieved April 22, 2009.

[72] "Crime Report for San Diego, California" (http://www.homesurfer.com/crimereports/view/crime_report.cfm?state=CA&area=San Diego). . Retrieved March 23, 2011.

[73] "City of San Diego website: Economic Development" (http://www.sandiego.gov/economic-development/sandiego/economy.shtml). Sandiego.gov. . Retrieved April 11, 2011.

[74] Powell, Ronald W. (October 17, 2007). "Tourism district OK'd by council" (http://www.signonsandiego.com/news/business/20071017-9999-1b17tourism.html). San Diego Union-Tribune. . Retrieved April 22, 2009.

[75] Eric Terrill; Julia Thomas, Anne Footer. "Submitted in response to Federal Funding Opportunity: FY 2011 Implementation of the U.S. Integrated Ocean Observing System (IOOS)" (http://sccoos.ucsd.edu/docs/FY11-16_IOOS_Proposal_web.pdf). *Southern California Coastal Ocean Observing System*. University of California San Diego. . Retrieved April 21, 2011.

[76] "Naval Base San Diego Thanks Navy League for Support" (http://www.navy.mil/search/display.asp?story_id=38356). U.S. Department of the Navy. . Retrieved April 7, 2011.

[77] "San Diego: the Birthplace of Naval Aviation Part One" (http://sandiegoairandspace.org/exhibits/naval_aviation_part_one/). San Diego Air & Space Museum. . Retrieved March 28, 2011.

[78] Kovach, Gretel C. and Kenney, Mary (June 15, 2011). "Carrier Carl Vinson returns home to San Diego" (http://www.signonsandiego.com/news/2011/jun/15/carl-vinson-returns-home-san-diego/). *Union Tribune*. . Retrieved June 20, 2011.

[79] "USS San Diego" (http://www.usssandiego.org/). San Diego Navy Historical Association. . Retrieved April 22, 2009.

[80] "Commission for Arts and Culture: Funding" (http://www.sandiego.gov/arts-culture/funding/index.shtml). City of San Diego. . Retrieved April 18, 2011.

[81] "Visitor Industry Summary" (http://www.sandiego.org/shared/file.download.php?id=394). San Diego Convention & Visitors Bureau. 2005 to 2009. . Retrieved April 7, 2011.

[82] "Summary of the Economic and fiscal Impact of Port Tidelands on the Region" (http://www.portofsandiego.org/public-documents/doc_view/738-economic-impact-fact-sheet.html). Unified Port of San Diego. 2007. . Retrieved Aug 31, 2011.

[83] Lori Weisberg (May 6, 2011). "San Diego losing another cruise ship" (http://www.signonsandiego.com/news/2011/may/06/cruises-mexican-riviera-likely-end-2012/). San Diego Union-Tribune. . Retrieved Aug 31, 2011.

[84] "Carnival Cruise Lines pulling out of San Diego" (http://www.signonsandiego.com/news/2011/jan/13/san-diego-to-lose-yet-another-cruise-ship/). *San Diego Union Tribune*. January 13, 2011. . Retrieved August 31, 2011.

[85] "City of San Diego:Foreign Trade Zone" (http://www.sandiego.gov/economic-development/sandiego/trade/tradezone.shtml). . Retrieved April 28, 2011.

[86] "Number of border crossings stabilizes" (http://www.signonsandiego.com/news/2010/jul/11/number-of-border-crossings-stabilizes/). *San Diego Union-Tribune*. July 11, 2010. . Retrieved April 28, 2011.

[87] "SANDAG: Otay Mesa Port of Entry Southbound Truck Route Improvements" (http://sandiegohealth.org/sandag/publicationid_853_1782.pdf). sandiegohealth.org. . Retrieved April 28, 2011.

[88] "Port of San Diego:10th Avenue Marine Terminal" (http://www.portofsandiego.org/maritime/tenth-avenue-terminal.html). . Retrieved April 28, 2011.

[89] "National ranking of California ports by cargo volume" (http://www.sddt.com/news/article.cfm?SourceCode=20110321czc). *San Diego Daily Transcript*. March 21, 2011. . Retrieved April 28, 2011.

[90] "iHub San Diego" (http://www.business.ca.gov/Portals/0/AdditionalResources/Reports/iHub Writeups-San Diego.pdf) (PDF). California Governor's Office of Economic Development. . Retrieved April 7, 2011.

[91] "City Of San Diego Largest Employers" (http://www.sddt.com/Databases/BusinessListings/ListCompanies.cfm?BusinessCategory_ID=140). San Diego Daily Transcript. . Retrieved April 22, 2009.

[92] Glazer, Joyce (October 6, 2008). "San Diego-based LG Mobile Phones donated $250,000 to the VH1 Save the Music Foundation" (http://www.entrepreneur.com/tradejournals/article/188738547.html). Entrepreneur Media. . Retrieved March 18, 2011.

[93] Desjardins, Doug (January 11, 2010). "Kyocera International to Get New Leader" (http://www.sdbj.com/news/2010/jan/11/kyocera-international-get-new-leader/). San Diego Business Journal. . Retrieved March 20, 2011.

[94] "Novatel website: Corporate headquarters" (http://www.novatelwireless.com/index.php?option=com_qcontacts&view=contact&id=3&Itemid=93). . Retrieved April 11, 2011.

[95] "Websense Named Top Software Company in San Diego County" (http://www.msnbc.msn.com/id/21661472/). MSNBC. November 6, 2007. . Retrieved April 22, 2009.

[96] Allen, Mike (September 20, 2010). "ESET Polishes the Apple, Now Protects Macs" (http://www.sdbj.com/news/2010/sep/20/eset-polishes-apple-now-protects-macs/). *San Diego Business Journal*. . Retrieved March 20, 2011.

[97] Doyle, Monica (February 5, 2004). "UCSD Extension Awarded A $150,000 Grant For Biotechnology Collaboration With Israel" (http://ucsdnews.ucsd.edu/newsrel/awards/US_Israel.asp). UCSD News. . Retrieved April 22, 2009.

[98] DeVol, Ross; Perry Wong, Junghoon Ki, Armen Bedroussian, and Rob Koepp (June 2004). "America's Biotech and Life Science Clusters: San Diego's Position and Economic Contributions" (http://www.milkeninstitute.org/publications/publications.taf?function=detail&ID=312&cat=ResRep). MilkenInstitute.org. . Retrieved April 22, 2009.

[99] "SDBN.org" (http://sdbn.org/directory/). SDBN.org. . Retrieved July 1, 2010.

[100] Walcott, Susan M. (5 2002). "Analyzing an Innovative Environment: San Diego as a Bioscience Beachhead" (http://edq.sagepub.com/content/16/2/99). *Economic Development Quarterly* **16** (2): 99–114. doi:10.1177/0891242402016002001. . Retrieved November 4, 2010.

[101] Bigelow, Bruce V. "San Diego's Life Sciences CROs—The Map of Clinical Research Organizations" (http://www.xconomy.com/san-diego/2010/01/27/san-diegos-life-sciences-cros-the-map-of-clinical-research-organizations/?single_page=true), "Xconomy", San Diego, January 27, 2010.

[102] California Association of Realtors (June 25, 2007). "C.A.R. Reports Sales Decrease 25 Percent in May, Median Price of a Home in California at $591,180, up 4.8 Percent from Year Ago" (http://www.businesswire.com/portal/site/google/index.

jsp?ndmViewId=news_view&newsId=20070625005991&newsLang=en). Business Wire. . Retrieved April 22, 2009.
[103] Weisberg, Lori (March 22, 2007). "Greener pastures outside of county?" (http://www.signonsandiego.com/uniontrib/20070322/news_1n22census.html). San Diego Union-Tribune. . Retrieved April 22, 2009.
[104] Freeman, Mike (December 29, 2010). "Housing Prices Fall Again, Index Says" (http://nl.newsbank.com/nl-search/we/Archives?p_action=doc&p_docid=134725AB44C9BD10&p_docnum=1&s_dlid=DL0111032600040931826&s_ecproduct=SUB-FREE&s_ecprodtype=INSTANT&s_trackval=&s_siteloc=&s_referrer=&s_subterm=Subscription until: 12/14/2025 11:59 PM&s_docsbal= &s_subexpires=12/14/2025 11:59 PM&s_docstart=&s_docsleft=&s_docsread=&s_username=sdubsub&s_accountid=AC0110122214325408110&s_upgradeable=no). *San Diego Union Tribune*. . Retrieved May 4, 2011.
[105] Showley, Roger (May 9, 2010). "Realty Revival" (http://nl.newsbank.com/nl-search/we/Archives?p_action=doc&p_docid=12FA5631673E3420&p_docnum=6&s_dlid=DL0111032600114313695&s_ecproduct=SUB-FREE&s_ecprodtype=INSTANT&s_trackval=&s_siteloc=&s_referrer=&s_subterm=Subscription until: 12/14/2025 11:59 PM&s_docsbal= &s_subexpires=12/14/2025 11:59 PM&s_docstart=&s_docsleft=&s_docsread=&s_username=sdubsub&s_accountid=AC0110122214325408110&s_upgradeable=no). *San Diego Union Tribune*. . Retrieved May 4, 2011.
[106] City of San Diego, California Comprehensive Annual Financial Report, for the Year ended June 30, 2009 (http://www.sandiego.gov/comptroller/reports/pdf/cafr_2009.pdf) Retrieved September 25, 2010
[107] "San Diego Unified School District – Our District" (http://www.sandi.net/20451072095932967/site/default.asp). San Diego Unified School District. . Retrieved May 31, 2011.
[108] Christie, Les (August 31, 2006). "America's smartest cities" (http://money.cnn.com/2006/08/29/real_estate/brainiest_cities/index.htm#list). CNNMoney.com. . Retrieved April 22, 2009.
[109] "Library Fact Sheet FY 2006" (http://www.sandiego.gov/public-library/about-the-library/libfactsheetfy07.shtml). San Diego Public Library. . Retrieved April 22, 2009.
[110] Hall, Matthew T. (April 12, 2006). "Budget spares libraries, parks" (http://www.signonsandiego.com/uniontrib/20060412/news_1m12preview.html). San Diego Union-Tribune. . Retrieved April 22, 2009.
[111] Gustafson, Craig (April 13, 2011). "Sanders proposes huge library, parks cuts" (http://www.signonsandiego.com/news/2011/apr/13/sanders-proposes-huge-library-park-cuts-to/). *Union-Tribune*. . Retrieved April 18, 2011.
[112] "San Diego Area Libraries" (http://infodome.sdsu.edu/research/libdirectory/index.shtml). San Diego State University. . Retrieved April 18, 2011.
[113] "La Jolla Playhouse" (http://www.lajollaplayhouse.org/about-the-playhouse). La Jolla Playhouse. . Retrieved July 1, 2010.
[114] "Old Globe Theater" (http://www.oldglobe.org/history/index.aspx). Oldglobe.org. December 2, 1937. . Retrieved July 1, 2010.
[115] "SoCal San Diego" (http://www.socal-sandiego.com/Movies/List_of_Movies_Filmed_in_SD.html). SoCal San Diego. . Retrieved March 12, 2011.
[116] "Journal of San Diego History, vol. 48, no. 2" (https://www.sandiegohistory.org/journal/2002-2/filming.htm). Sandiegohistory.org. . Retrieved March 12, 2011.
[117] "OMBAC Rugby Home" (http://www.ombac.org/ombac_rugby/). Ombac.org. . Retrieved July 1, 2010.
[118] "About" (http://www.toddclever.com/about). Todd Clever. January 16, 1983. . Retrieved July 1, 2010.
[119] "RL Hopes to Move West" (http://www.americanrugbynews.com/artman/publish/rugby_league/RL_Hopes_To_Move_West.shtml). Americanrugbynews.com. . Retrieved March 12, 2011.
[120] Granberry, Mike (July 10, 1981). "Over-the-Line" (http://pqasb.pqarchiver.com/latimes/access/658715592.html?dids=658715592:658715592&FMT=ABS&FMTS=ABS:AI&type=historic&date=Jul+10,+1981&author=&pub=Los+Angeles+Times&desc=Over-the-Line&pqatl=google). *Los Angeles Times*. . Retrieved May 4, 2011.
[121] "Over-the-Line official website" (http://www.ombac.org/over_the_line/). Ombac.org. . Retrieved May 4, 2011.
[122] "San Diego News, Local, California and National News" (http://www.signonsandiego.com). SignOnSanDiego.com. . Retrieved March 12, 2011.
[123] "San Diego, Phoenix and Detroit Lead Broadband Wired Cities, According to Nielsen//NetRatings" (http://www.nielsen-netratings.com/pr/pr_040915.pdf) (PDF). Nielsen//NetRatings. September 15, 2004. . Retrieved April 25, 2011.
[124] Stigall, Gary (May 3, 1999). "KFMB-TV Turns 50" (http://www.sbe36.org/1999/0509_kfmbtv50.html). Society of Broadcast Engineers Chapter 36 San Diego. . Retrieved April 22, 2009.
[125] San Diego market in "Market Profiles" (http://www.tvb.org/market_profiles). Television Bureau of Advertising. . Retrieved April 25, 2011.
[126] "San Diego City website" (http://www.sandiego.gov/mayortransition/index.shtml). Sandiego.gov. . Retrieved July 1, 2010.
[127] "San Diego City website" (http://www.sandiego.gov/city-clerk/elections/city/details.shtml). Sandiego.gov. . Retrieved July 1, 2010.
[128] Pollack, Andrew (November 9, 2005). "Republican Wins San Diego Mayor's Race" (http://www.nytimes.com/2005/11/09/national/09cnd-sandiego.html). *The New York Times*. . Retrieved April 6, 2011.
[129] "Voter Registration in the City of San Diego" (http://www.sandiego.gov/city-clerk/pdf/voterstats.pdf) (PDF). San Diego Office of the City Clerk (http://www.sandiego.gov/city-clerk/). August 1, 2007. . Retrieved April 22, 2009.
[130] "Registrar of voters: Maps of individual supervisorial districts" (http://www.sdcounty.ca.gov/voters/Eng/Ehandoutmap.shtml). County of San Diego. . Retrieved May 31, 2011.
[131] "San Diego County website" (http://www.sdcounty.ca.gov/general/bos.html). sdcounty.ca.gov. . Retrieved December 14, 2010.

[132] Hinckley, Catie; Walker, John (November 1, 2006). "Will Gerrymandered Districts Stem the Wave of Voter Unrest?" (http://www.clcblog.org/blog_item-85.html). Campaign Legal Center. . Retrieved April 22, 2009.

[133] Horstman, Barry (December 6, 1987). "Man About Town : San Diego's Ex-Mayor Roger Hedgecock Hasn't Let His Felony Conviction Get Him Down. But This Week, the Past May Catch Up With Him." (http://articles.latimes.com/1987-12-06/magazine/tm-27105_1_san-diego-beach). *The Los Angeles Times*. . Retrieved April 2, 2011.

[134] Abrahamson, Alan (February 2, 1992). "Bailiff's Bias in Hedgecock Trial Disclosed" (http://articles.latimes.com/1992-02-02/news/mn-1802_1_fair-trial). *The Los Angeles Times*. . Retrieved April 3, 2011.

[135] "Hedgecock has clean slate; judge erases felony record" (http://nl.newsbank.com/nl-search/we/Archives?p_action=doc&p_docid=11782F5483FF36EF&p_docnum=4&s_dlid=DL0111060115504727581&s_ecproduct=SUB-FREE&s_ecprodtype=INSTANT&s_trackval=&s_siteloc=&s_referrer=&s_subterm=Subscription until: 12/14/2025 11:59 PM&s_docsbal= &s_subexpires=12/14/2025 11:59 PM&s_docstart=&s_docsleft=&s_docsread=&s_username=sdubsub&s_accountid=AC0110122214325408110&s_upgradeable=no). *San Diego Union-Tribune*. January 1, 1991. . Retrieved June 1, 2011.

[136] "San Diego's Widening Pension Woes" (http://www.businessweek.com/magazine/content/05_24/b3937087.htm). *Bloomberg BusinessWeek*. June 13, 2005. . Retrieved July 1, 2010.

[137] Strumpf, Daniel (June 15, 2005) San Diego's Pension Scandal for Dummies (http://replay.waybackmachine.org/20090219224628/http://www.sdcitybeat.com/cms/story/detail/?id=3244), San Diego City Beat via Internet Archive. Retrieved on April 3, 2011.

[138] Hall, Matthew T. (April 8, 2010). "Five cleared in San Diego pension case" (http://www.signonsandiego.com/news/2010/apr/08/five-cleared-in-pension-case/). *San Diego Union-Tribune*. . Retrieved July 1, 2010.

[139] "Cunningham Moving to Arizona Prison" (http://www.washingtonpost.com/wp-dyn/content/article/2007/01/05/AR2007010501858.html). *Washington Post*. January 5, 2007. . Retrieved February 3, 2010.

[140] Greg Moran and Kelly Thornton (July 19, 2005). "*Councilmen Guilty*" (http://nl.newsbank.com/nl-search/we/Archives?p_action=doc&p_docid=10B7E53625734BA8&p_docnum=1&s_dlid=DL0111040622315622760&s_ecproduct=SUB-FREE&s_ecprodtype=INSTANT&s_trackval=&s_siteloc=&s_referrer=&s_subterm=Subscription until: 12/14/2025 11:59 PM&s_docsbal= &s_subexpires=12/14/2025 11:59 PM&s_docstart=&s_docsleft=&s_docsread=&s_username=sdubsub&s_accountid=AC0110122214325408110&s_upgradeable=no). *San Diego Union-Tribune*. . Retrieved April 6, 2011.

[141] "Appeals Court opinion, Sept. 1, 2009" (http://www.ca9.uscourts.gov/datastore/opinions/2009/09/01/05-50902.pdf) (PDF). . Retrieved July 1, 2010.

[142] Greg Moran (October 14, 2010). "*Seven Years Later, Zucchet Cleared*" (http://nl.newsbank.com/nl-search/we/Archives?p_action=doc&p_docid=132E15958E125350&p_docnum=3&s_dlid=DL0111040622192220299&s_ecproduct=SUB-FREE&s_ecprodtype=INSTANT&s_trackval=&s_siteloc=&s_referrer=&s_subterm=Subscription until: 12/14/2025 11:59 PM&s_docsbal= &s_subexpires=12/14/2025 11:59 PM&s_docstart=&s_docsleft=&s_docsread=&s_username=sdubsub&s_accountid=AC0110122214325408110&s_upgradeable=no). *San Diego Union-Tribune*. . Retrieved April 6, 2011.

[143] Marshall, David. San Diego's Balboa Park (http://books.google.com/books?id=tG3asbfLcUsC&pg=PA110&dq=163+beautiful+diego&hl=en&ei=23aOTagzyZyBB_DtxK8N&sa=X&oi=book_result&ct=result&resnum=9&ved=0CFEQ6AEwCA#v=onepage&q=163 beautiful diego&f=false). Arcadia Publishing. 2007.

[144] "San Diego, Madison (WI) and Boulder (CO) Are Best among Cities of Their Size, While Atlanta, Boston and Houston Are Worst" (http://www.bikechattanooga.org/BicyclingMagazineRecognizesChattanoogainTop21Cities.html). Bicycling. January 26, 2006. . Retrieved April 22, 2009.

[145] "Transit.sd511.com" (http://www.sdcommute.com/). . and "Metropolitan Transit System" (http://www.sdmts.com/). SANDAG. . Retrieved April 18, 2011.

[146] "511 Overview" (http://www.511sd.com/About511.aspx). SANDAG. . Retrieved April 18, 2011.

[147] Downey, Dave (April 24, 2006). "FAA chief says region right to consider bases" (http://www.nctimes.com/articles/2006/04/25/news/top_stories/20_02_594_24_06.txt). *North County Times*. . Retrieved April 22, 2009.

[148] "City of San Diego:Airports" (http://www.sandiego.gov/airports/). Sandiego.gov. . Retrieved May 4, 2011.

[149] Hall, Matthew (May 2, 2007). "City: 37 percent of streets in acceptable driving condition" (http://www.signonsandiego.com/news/metro/20070502-1610-bn02streets.html). San Diego Union-Tribune. . Retrieved April 22, 2009.

[150] "San Diego International Airport" (http://www.san.org/). San.org. . Retrieved May 4, 2011.

[151] "2011 City and Neighborhood Rankings" (http://www.walkscore.com/rankings/cities/). Walk Score. 2011. . Retrieved Aug 28, 2011.

[152] "Online Directory: California, USA" (http://web.archive.org/web/20080116164532/http://www.sister-cities.org/icrc/directory/usa/CA). Sister Cities International. Archived from the original (http://www.sister-cities.org/icrc/directory/usa/CA) on January 16, 2008. . Retrieved April 22, 2009.

[153] **(Chinese (PRC))** "" (http://www.fjfao.gov.cn/index/noDateCategory?id=51) (in zh-hans). . . Retrieved 2008-03-05.

[154] "Miasta partnerskie Warszawy" (http://um.warszawa.pl/v_syrenka/new/index.php?dzial=aktualnosci&ak_id=3284&kat=11). *um.warszawa.pl*. Biuro Promocji Miasta. May 4, 2005. . Retrieved August 29, 2008.

[155] "Eight Cities/Six Ports: Yokohama's Sister Cities/Sister Ports" (http://www.welcome.city.yokohama.jp/eng/tourism/mame/a3000.html). Yokohama Convention & Visitiors Bureau (http://www.welcome.city.yokohama.jp/eng/ycvb/index.html). . Retrieved July 18, 2009.

External links

- City of San Diego Official Website (http://www.sandiego.gov/)
- City of San Diego Redevelopment Agency Website (http://www.sandiego.gov/redevelopment-agency)
- Centre City Development Corporation Website (http://www.ccdc.com/)
- Southeastern Economic Development Corporation Website (http://www.sedcinc.com/)
- SANDAG, San Diego's Regional Planning Agency (http://www.sandag.org/)
- Demographic Fact Sheet (http://quickfacts.census.gov/qfd/states/06/0666000.html) from Census Bureau
- History of San Diego (http://sandiegohistory.org/index.html) from San Diego Historical Society (http://sandiegohistory.org/index.html)
- San Diego Unified School District (http://www.sandi.net/sandi/site/default.asp)
- San Diego Public Library (http://www.sandiegolibrary.org/)
- San Diego Convention and Visitors Bureau (http://www.sandiego.org/)
- San Diego Wiki
- San Diego travel guide from Wikitravel

Groundwater

Shipot, a common source of drinking water in a Ukrainian village.

Groundwater is water located beneath the ground surface in soil pore spaces and in the fractures of rock formations. A unit of rock or an unconsolidated deposit is called an aquifer when it can yield a usable quantity of water. The depth at which soil pore spaces or fractures and voids in rock become completely saturated with water is called the water table. Groundwater is recharged from, and eventually flows to, the surface naturally; natural discharge often occurs at springs and seeps, and can form oases or wetlands. Groundwater is also often withdrawn for agricultural, municipal and industrial use by constructing and operating extraction wells. The study of the distribution and movement of groundwater is hydrogeology, also called groundwater hydrology.

Typically, groundwater is thought of as liquid water flowing through shallow aquifers, but technically it can also include soil moisture, permafrost (frozen soil), immobile water in very low permeability bedrock, and deep geothermal or oil formation water. Groundwater is hypothesized to provide lubrication that can possibly influence the movement of faults. It is likely that much of the Earth's subsurface contains some water, which may be mixed with other fluids in some instances. Groundwater may not be confined only to the Earth. The formation of some of the landforms observed on Mars may have been influenced by groundwater. There is also evidence that liquid water may also exist in the subsurface of Jupiter's moon Europa.[1]

The entire surface water flow of the Alapaha River near Jennings, Florida going into a sinkhole leading to the Floridan Aquifer groundwater

Aquifers

An *aquifer* is a layer of porous substrate that contains and transmits groundwater. When water can flow directly between the surface and the saturated zone of an aquifer, the aquifer is unconfined. The deeper parts of unconfined aquifers are usually more saturated since gravity causes water to flow downward.

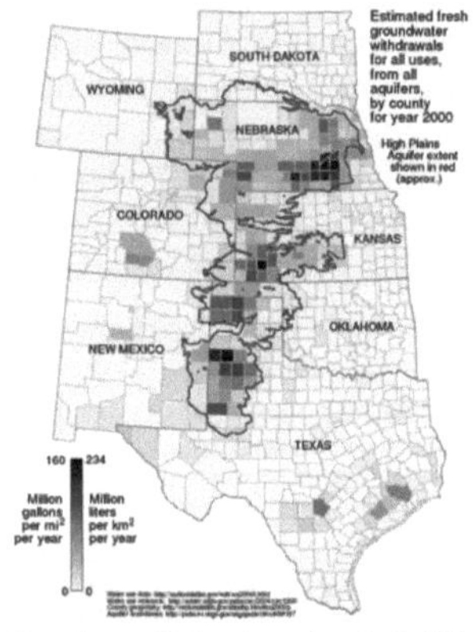

Groundwater withdrawal rates from the Ogallala Aquifer in the central U.S.

The upper level of this saturated layer of an unconfined aquifer is called the *water table* or *phreatic surface*. Below the water table, where generally all pore spaces are saturated with water is the phreatic zone.

Substrate with low porosity that permits limited transmission of groundwater is known as an *aquitard*. An *aquiclude* is a substrate with porosity that is so low it is virtually impermeable to groundwater.

A *confined aquifer* is an aquifer that is overlain by a relatively impermeable layer of rock or substrate such as an aquiclude or aquitard. If a confined aquifer follows a downward grade from its *recharge zone*, groundwater can become pressurized as it flows. This can create artesian wells that flow freely without the need of a pump and rise to a higher elevation than the static water table at the above, unconfined, aquifer.

The characteristics of aquifers vary with the geology and structure of the substrate and topography in which they occur. Generally, the more productive aquifers occur in sedimentary geologic formations. By comparison, weathered and fractured crystalline rocks yield smaller quantities of groundwater in many environments. Unconsolidated to poorly cemented alluvial materials that have accumulated as valley-filling sediments in major river valleys and geologically subsiding structural basins are included among the most productive sources of groundwater.

The high specific heat capacity of water and the insulating effect of soil and rock can mitigate the effects of climate and maintain groundwater at a relatively steady temperature. In some places where groundwater temperatures are maintained by this effect at about 50°F/10°C, groundwater can be used for controlling the temperature inside structures at the surface. For example, during hot weather relatively cool groundwater can be pumped through radiators in a home and then returned to the ground in another well. During cold seasons, because it is relatively warm, the water can be used in the same way as a source of heat for heat pumps that is much more efficient than using air.

Water cycle

Groundwater makes up about twenty percent of the world's fresh water supply, which is about 0.61% of the entire world's water, including oceans and permanent ice. Global groundwater storage is roughly equal to the total amount of freshwater stored in the snow and ice pack, including the north and south poles. This makes it an important resource which can act as a natural storage that can buffer against shortages of surface water, as in during times of drought.[2]

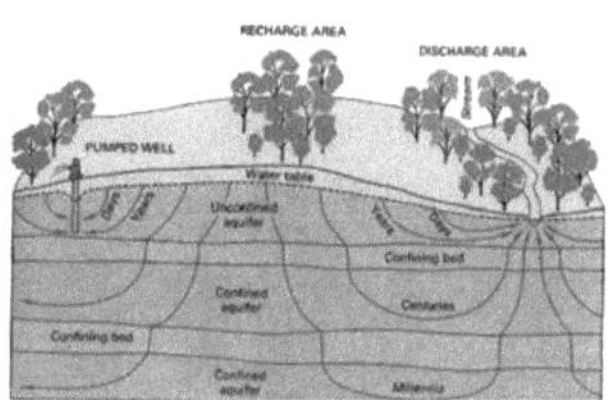
Relative groundwater travel times.

Groundwater is naturally replenished by surface water from precipitation, streams, and rivers when this recharge reaches the water table.

Groundwater can be a long-term 'reservoir' of the natural water cycle (with residence times from days to millennia), as opposed to short-term water reservoirs like the atmosphere and fresh surface water (which have residence times from minutes to years). The figure shows how deep groundwater (which is quite distant from the surface recharge) can take a very long time to complete its natural cycle.

The Great Artesian Basin in central and eastern Australia is one of the largest confined aquifer systems in the world, extending for almost 2 million km^2. By analysing the trace elements in water sourced from deep underground, hydrogeologists have been able to determine that water extracted from these aquifers can be more than 1 million years old.

By comparing the age of groundwater obtained from different parts of the Great Artesian Basin, hydrogeologists have found it increases in age across the basin. Where water recharges the aquifers along the Eastern Divide, ages are young. As groundwater flows westward across the continent, it increases in age, with the oldest groundwater occurring in the western parts. This means that in order to have travelled almost 1000 km from the source of recharge in 1 million years, the groundwater flowing through the Great Artesian Basin travels at an average rate of about 1 metre per year.

Issues

Overview

Certain problems have beset the use of groundwater around the world. Just as river waters have been over-used and polluted in many parts of the world, so too have aquifers. The big difference is that aquifers are out of sight. The other major problem is that water management agencies, when calculating the 'sustainable yield' of aquifer and river water, have often counted the same water twice, once in the aquifer, and once in its connected river. This problem, although understood for centuries, has persisted, partly through inertia within government agencies. In Australia, for example, prior to the statutory reforms initiated by the Council of Australian Governments water reform framework in the 1990s, many Australian States managed groundwater and surface water through separate government agencies, an approach beset by rivalry and poor communication.

The time lags inherent in the dynamic response of groundwater to development have generally been ignored by water management agencies, decades after scientific understanding of the issue was consolidated. In brief, the effects of groundwater overdraft (although undeniably real) may take decades or centuries to manifest themselves. In a classic study in 1982, Bredehoeft and colleagues[3] modelled a situation where groundwater extraction in an intermontane basin withdrew the entire annual recharge, leaving 'nothing' for the natural groundwater-dependent vegetation community. Even when the borefield was situated close to the vegetation, 30% of the original vegetation demand could still be met by the lag inherent in the system after 100 years. By year 500 this had reduced to 0%, signalling complete death of the groundwater-dependent vegetation. The science has been available to make these calculations for decades; however water management agencies have generally ignored effects which will appear outside the rough timeframe of political elections (3 to 5 years). Marios Sophocleous[3] argued strongly that management agencies must define and use appropriate timeframes in groundwater planning. This will mean calculating groundwater withdrawal permits based on predicted effects decades, sometimes centuries in the future.

As water moves through the landscape it collects soluble salts, mainly sodium chloride. Where such water enters the atmosphere through evapotranspiration, these salts are left behind. In irrigation districts, poor drainage of soils and surface aquifers can result in water tables coming to the surface in low-lying areas. Major land degradation problems of soil salinity and waterlogging result,[4] combined with increasing levels of salt in surface waters. As a consequence, major damage has occurred to local economies and environments.[5]

Four important effects are worthy of brief mention. First, flood mitigation schemes, intended to protect infrastructure built on floodplains, have had the unintended consequence of reducing aquifer recharge associated with natural flooding. Second, prolonged depletion of groundwater in extensive aquifers can result in land subsidence, with associated infrastructure damage – as well as (thirdly) saline intrusion.[6] Fourth, draining acid sulphate soils, often found in low-lying coastal plains, can result in acidification and pollution of formerly freshwater and estuarine streams.[7]

Another cause for concern is that groundwater drawdown from over-allocated aquifers has the potential to cause severe damage to both terrestrial and aquatic ecosystems – in some cases very conspicuously but in others quite imperceptibly because of the extended period over which the damage occurs.[8]

Overdraft

Groundwater is a highly useful and often abundant resource. However, over-use, or overdraft, can cause major problems to human users and to the environment. The most evident problem (as far as human groundwater use is concerned) is a lowering of the water table beyond the reach of existing wells. Wells must consequently be deepened to reach the groundwater; in some places (e.g., California, Texas and India) the water table has dropped hundreds of feet because of extensive well pumping. In the Punjab region of India, for example, groundwater levels have dropped 10 meters since 1979, and the rate of depletion is accelerating.[9] A lowered water table may, in turn, cause other problems such as groundwater-related subsidence and saltwater intrusion.

Wetlands contrast the arid landscape around Middle Spring, Fish Springs National Wildlife Refuge, Utah.

Groundwater is also ecologically important. The importance of groundwater to ecosystems is often overlooked, even by freshwater biologists and ecologists. Groundwaters sustain rivers, wetlands and lakes, as well as subterranean ecosystems within karst or alluvial aquifers.

Not all ecosystems need groundwater, of course. Some terrestrial ecosystems – for example, those of the open deserts and similar arid environments – exist on irregular rainfall and the moisture it delivers to the soil, supplemented by moisture in the air. While there are other terrestrial ecosystems in more hospitable environments where groundwater plays no central role, groundwater is in fact fundamental to many of the world's major ecosystems. Water flows between groundwaters and surface waters. Most rivers, lakes and wetlands are fed by, and (at other places or times) feed groundwater, to varying degrees. Groundwater feeds soil moisture through percolation, and many terrestrial vegetation communities depend directly on either groundwater or the percolated soil moisture above the aquifer for at least part of each year. Hyporheic zones (the mixing zone of streamwater and groundwater) and riparian zones are examples of ecotones largely or totally dependent on groundwater.

Aquifer drawdown or overdrafting and the pumping of fossil water increases the total amount of water within the hydrosphere subject to transpiration and evaporation processes, thereby causing accretion in water vapour and cloud cover, the primary absorbers of infrared radiation in the Earth's atmosphere. Adding water to the system has a forcing effect on the whole earth system, an accurate estimate of which hydrogeological fact is yet to be quantified.

A solution to over-use of groundwater

Irrigated agriculture is a big business in Asia. It accounts for 70 per cent of the world's irrigated land and 73 per cent of water used each year by the farming industry. As ageing large-scale surface irrigation schemes have become increasingly inefficient, and farmers have begun growing a wider range of crops requiring water on demand, the number of groundwater wells in India has exploded.[10] In 1960, there were fewer than 100,000 such wells; by 2006 the figure had risen to nearly 12 million.[11] In India, a possible solution to over-use of groundwater is emerging, known as 'groundwater recharge'. It involves capturing rainwater that would otherwise run-off, and using it to refill aquifers. Since 2000, the International Water Management Institute (IWMI) has been working with the Indian authorities to help improve the availability of water for agriculture in India. In 2006, India's finance minister invited IWMI to submit policy recommendations based on its research on groundwater depletion. One of the key recommendations was to instigate a programme of recharging groundwater across the 65 per cent of India that has hard-rock aquifers. As a result, the Indian government allocated Rs 1800 crore (US$400million) to fund dug-well recharge projects (a dug-well is a wide, shallow well, often lined with concrete) in 100 districts within seven states where water stored in hard-rock aquifers has been over-exploited. These geological formations have a much lower capacity to store rainwater than alluvial areas with porous sand or clay rocks, hence being given priority. The money is sufficient to fund seven million structures to be installed on dug-wells to divert monsoon runoff. The structures include a de-siltation chamber, plus pipes to collect surplus rainwater and divert de-silted water from the chamber to the well.[12] As of the end of November 2009, funds amounting to Rs. 216.98 crore (including Rs. 199.98 crore as subsidy to beneficiaries and 17 crore for IEC/Capacity Building activities) had been released to the concerned states. Subsidies had been released to 566,637 beneficiaries.[13]

Subsidence

Subsidence occurs when too much water is pumped out from underground, deflating the space below the above-surface, and thus causing the ground to actually collapse. The result can look like craters on plots of land. This occurs because in its natural equilibrium state, the hydraulic pressure of groundwater in the pore spaces of the aquifer and the aquitard supports some of the weight of the overlying sediments. When groundwater is removed from aquifers by excessive pumping, pore pressures in the aquifer drop and compression of the aquifer may occur. This compression may be partially recoverable if pressures rebound, but much of it is not. When the aquifer gets compressed it may cause land subsidence, a drop in the ground surface. The city of New Orleans, Louisiana, is actually below sea level today, and its subsidence is partly caused by removal of groundwater from the various aquifer/aquitard systems beneath it. In the first half of the 20th century, the city of San Jose, California, dropped 13 feet from land subsidence caused by overpumping; this subsidence has been halted with improved groundwater management.

Seawater intrusion

Generally, in very humid or undeveloped regions, the shape of the water table mimics the slope of the surface. The recharge zone of an aquifer near the seacoast is likely to be inland, often at considerable distance. In these coastal areas, a lowered water table may induce sea water to reverse the flow toward the land. Sea water moving inland is called a saltwater intrusion. Alternatively, salt from mineral beds may leach into the groundwater of its own accord.

Mining

Sometimes the water movement from the recharge zone to the place where it is withdrawn may take centuries (see figure above). When the usage of water is greater than the recharge, it is referred to as *mining* water (the water is often called fossil water because of its geologic age). Under those circumstances it is not a renewable resource.

Pollution

Water pollution of groundwater, from pollutants released to the ground that can work their way down into groundwater, can create a contaminant plume within an aquifer. Movement of water and dispersion within the aquifer spreads the pollutant over a wider area, its advancing boundary often called a plume edge, which can then intersect with groundwater wells or daylight into surface water such as seeps and springs, making the water supplies unsafe for humans and wildlife. The interaction of groundwater contamination with surface waters is analyzed by use of hydrology transport models.

Iron oxide staining caused by reticulation from an unconfined aquifer in karst topography. Perth, Western Australia.

The stratigraphy of the area plays an important role in the transport of these pollutants. An area can have layers of sandy soil, fractured bedrock, clay, or hardpan. Areas of karst topography on limestone bedrock are sometimes vulnerable to surface pollution from groundwater. Earthquake faults can also be entry routes for downward contaminant entry. Water table conditions are of great importance for drinking water supplies, agricultural irrigation, waste disposal (including nuclear waste), wildlife habitat, and other ecological issues [14] .

In the US, upon commercial real estate property transactions both groundwater and soil are the subjects of scrutiny, with a Phase I Environmental Site Assessment normally being prepared to investigate and disclose potential pollution issues[15] . In the San Fernando Valley of California, Real estate contracts for property transfer below the Santa Susana Field Laboratory (SSFL) and eastward have clauses releasing the seller from liability for groundwater contamination consequences from existing or future water pollution of the Valley Aquifer.

Love Canal was one of the most widely known examples of groundwater pollution. In 1978, residents of the Love Canal neighbourhood in upstate New York noticed high rates of cancer and an alarming number of birth defects. This was eventually traced to organic solvents and dioxins from an industrial landfill that the neighbourhood had been built over and around, which had then infiltrated into the water supply and evaporated in basements to further contaminate the air. Eight hundred families were reimbursed for their homes and moved, after extensive legal battles and media coverage.

Another example of widespread groundwater pollution is in the Ganges Plain of northern India and Bangladesh where severe contamination of groundwater by naturally occurring arsenic affects 25% of water wells in the shallower of two regional aquifers. The pollution occurs because aquifer sediments contain organic matter (dead plant material) that generates anaerobic (an environment without oxygen) conditions in the aquifer. These conditions result in the microbial dissolution of iron oxides in the sediment and thus the release of the arsenic, normally strongly bound to iron oxides, into the water. As a consequence, arsenic-rich groundwater is often iron-rich, although secondary processes often obscure the association of dissolved arsenic and dissolved iron.

Ground Water Rule

In November 2006, the Environmental Protection Agency published the Ground Water Rule in the United States Federal Register. The EPA was worried that the ground water system would be vulnerable to contamination from fecal matter. The point of the rule was to keep microbial pathogens out of public water sources [16] The 2006 Ground Water Rule was an amendment of the 1996 Safe Drinking Water Act.

See also

- Groundwater models
- Groundwater flow
- Baseflow
- Seep (hydrology)
- Spring (hydrosphere)
- Water well
- Optimum water content for tillage

References

[1] Richard Greenburg (2005). *The Ocean Moon: Search for an Alien Biosphere*. Springer Praxis Books.

[2] "Learn More: Groundwater" (http://water.columbia.edu/?id=learn_more&navid=groundwater/). Columbia Water Center. . Retrieved 2009-09-15.

[3] Sophocleous, Marios (2002). "Interactions between groundwater and surface water: the state of the science". *Hydrogeology Journal* **10**: 52–67. doi:10.1007/s10040-001-0170-8.

[4] "Free articles and software on drainage of waterlogged land and soil salinity control" (http://www.waterlog.info). . Retrieved 2010-07-28.

[5] Ludwig, D.; Hilborn, R.; Walters, C. (1993). "Uncertainty, Resource Exploitation, and Conservation: Lessons from History" (http://landscape.forest.wisc.edu/courses/Landscape565spr01/Ludwig_etal1993.pdf). *Science* **260** (5104): 17–36. doi:10.1126/science.260.5104.17. JSTOR 1942074. PMID 17793516. .

[6] Zektser et al.

[7] Sommer, Bea; Horwitz, Pierre; Sommer, Bea; Horwitz, Pierre (2001). "Water quality and macroinvertebrate response to acidification following intensified summer droughts in a Western Australian wetland". *Marine and Freshwater Research* **52** (7): 1015. doi:10.1071/MF00021.

[8] Zektser, S.; Lo�Iciga, H. A.; Wolf, J. T. (2004). "Environmental impacts of groundwater overdraft: selected case studies in the southwestern United States". *Environmental Geology* **47** (3): 396–404. doi:10.1007/s00254-004-1164-3.

[9] Upmanu Lall. "Punjab: A tale of prosperity and decline" (http://blogs.ei.columbia.edu/water/2009/07/28/punjab-a-tale-of-prosperity-and-decline/). Columbia Water Center. . Retrieved 2009-09-11.

[10] Mukherji, A. Revitalising Asia's Irrigation: To sustainably meet tomorrow's food needs (http://www.iwmi.cgiar.org/Publications/Other/index.aspx) 2009, IWMI and FAO

[11] Chapter 10: The Groundwater Recharge Movement in India (http://www.iwmi.cgiar.org/publications/CABI_Publications/CA_CABI_Series/Ground_Water/protected/Giordano_1845931726-Chapter10.pdf), By Sakthivadivel, R. in *The Agricultural Groundwater Revolution*, Ed. Giordano, M. and Villholth, K, CABI Publications, 2007

[12] Influencing irrigation policy in India (http://www.iwmi.cgiar.org/Publications/Success_Stories/index.aspx), Success Story 6, 2010, International Water Management Institute, Colombo, Sri Lanka

[13] Initiatives in Water Resources (http://indiacurrentaffairs.org/initiatives-in-water-resources), Indian Current Affairs, December 30, 2009, page accessed 5 May 2011

[14] Groundwater Sampling; http://www.groundwatersampling.org/

[15] EPA; http://water.epa.gov/type/groundwater/index

[16] Ground Water Rule (GWR) | Ground Water Rule | US EPA (http://water.epa.gov/lawsregs/rulesregs/sdwa/gwr/index.cfm). Water.epa.gov. Retrieved on 2011-06-09.

External links

- USGS Office of Groundwater (http://water.usgs.gov/ogw/)
- UK Groundwater Forum (http://www.groundwateruk.org)
- Status of Ground Water in India (http://www.indiawaterportal.org/channels/groundwater)
- The Groundwater Foundation (http://www.groundwater.org)
- Groundwater Information from the Coastal Ocean Institute (http://www.whoi.edu/page.do?pid=11929)
- US National Ground Water Association (http://www.ngwa.org)
- Connected Waters Initiative, University of New South Wales (http://www.connectedwaters.unsw.edu.au)
- American Water Resources Association (http://www.awra.org)
- Multimedia course on groundwater (http://www.indiawaterportal.org/multimedia.html)
- Bibliography on Water Resources and International Law (http://www.ppl.nl/index.php?option=com_wrapper&view=wrapper&Itemid=82) Peace Palace Library
- IGRAC, International Groundwater Resources Assessment Centre (http://www.igrac.net)
- Groundwater (http://www.hydrology.nl/key-publications/231-groundwater-geology-of-the-netherlands.html) Chapter on groundwater in the Netherlands, from 'Geology of the Netherlands'
- Arie S.Issar. The Evolution of Groundwater Exploitation Methods in the Middle East through History. (http://www.chaire-rome.hst.ulaval.ca/Docs_revue/html/revue_etat_questions_arie_issar_evolution_groundwater_exploitation_methods.htm)
- US Army Geospatial Center (http://www.agc.army.mil/) — For information on OCONUS surface water and groundwater.

San Luis Rey River

San Luis Rey River	
River	
Mouth of San Luis Rey River 06-27-2003 10:30 AM Mouth of river from N Coast Hwy	
Country	United States
State	California
Tributaries	
- left	Buena Vista Creek
- right	West Fork San Luis Rey River, Pauma Creek
Source	Lake Henshaw
- location	Confluence of West Fork San Luis Rey River and Buena Vista Creek, Cleveland National Forest, San Diego County
- elevation	2627 ft (801 m)
- coordinates	33°24′02″N 116°37′26″W [1]
Mouth	Pacific Ocean
- location	Oceanside, San Diego County
- elevation	0 ft (0 m)
- coordinates	33°12′08″N 117°23′32″W [1]
Length	69 mi (111 km), East-west [2]
Basin	557 sq mi (1443 km^2) [3]
Discharge	for Oceanside
- average	36.3 cu ft/s (1 m^3/s) [3]
- max	25700 cu ft/s (728 m^3/s)
- min	0 cu ft/s (0 m^3/s)

The **San Luis Rey River** is a river in northern San Diego County, California. The river's headwaters are in the Cleveland National Forest near Palomar Mountain. The river drains into the Pacific Ocean on the northern end of the city of Oceanside. The river is over 69 miles (111 km) long[4] and drains 562 square miles (1460 km^2). There is little water in the river during most of the year, but it can have very large flows during winter storms. The dam that forms Lake Henshaw is the only one on the river itself, but the river's drainage is extensively dammed. During the last 7 miles (11 km), the river flows through a 400-foot-wide (120 m) earthen channel with levees on each side to prevent flooding, although the rest of the river remains unchannelized. Urbanization, mining and agriculture have caused substantial deterioration of the watershed. Lake Henshaw, relatively close to the headwaters, is a source of drinking water. However, the water in the lower part of the river is only used for groundwater recharge. In 2009, a bridge was completed across the river near its mouth, replacing a small causeway that used to wash away every few years.

Course

The San Luis Rey River rises in two main branches. The mainstem starts east of Rocky Mountain in the Cleveland National Forest and flows generally south-southwest. The West Fork's headwaters rise as a pair of tiny streams, Fry Creek and Iron Springs Creek, just to the north of Palomar Mountain. These two streams combine into the West Fork, which flows southeast through the Mendenhall Valley. The West Fork joins the main stem at Lake Henshaw, a reservoir formed by a dam across the main stem San Luis Rey River.

From the base of Henshaw Dam, the river winds west along the foot of the Palomars, followed by California State Route 76. It then bends southwest into a gorge. It leaves the canyon at the wide and spacious Pauma Valley, where it becomes a wash surrounded on both sides by agricultural fields. Potrero and Pauma Creeks enter from the right, then Frey Creek and Agua Tibia Creek as the river travels northwest. The river turns west, passing Pala, flowing through a patchwork of privately owned, government-owned and Native American lands. It then crosses under Interstate 15 and exits the foothills of the mountains near Bonsall.

After passing Bonsall the river flows generally southwest, through the cities of San Luis Rey and Oceanside. Although usually dry up to this point the river starts to contain water as it nears the mouth, both from seeps and from tidal activity that forms a lagoon at the mouth of the river. The San Luis Rey empties into the Pacific north of Oceanside, about 2 miles (3.2 km) south of the Santa Margarita River and ~30 miles (48 km) north of San Diego.

In popular culture

The 2010 Rockstar Games release "Red Dead Redemption" features a river inspired by the San Luis Rey. The "San Luis River" serves as the border between the Western Border States and Mexico and is one of the hallmarks of the game.

Gallery

Pacific St and mouth of river

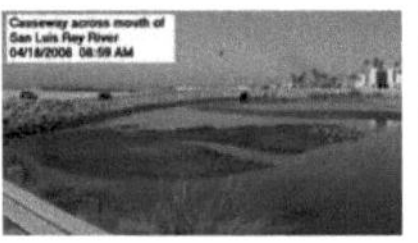

Former causeway across mouth of river

Lake Henshaw from Mesa Grande Rd

Lake Henshaw from road to Love Valley

Lake Henshaw from lookout on County Highway S7

Underneath the Benet Rd bridge after several days of intense rainfall in winter 2010

References

[1] "San Luis Rey River" (http://geonames.usgs.gov/pls/gnispublic/f?p=gnispq:3:::NO::P3_FID:273500). Geographic Names Information System, U.S. Geological Survey. 1981-01-19. . Retrieved 2010-09-12.

[2] Length measured in Google Earth using path measure tool; measured to longest source (includes Fry Creek and West Fork San Luis Rey River

[3] "Water-Data Report 2009: USGS Gage #11042000 on the San Luis Rey River at Oceanside, CA" (http://wdr.water.usgs.gov/wy2009/pdfs/11042000.2009.pdf). *National Water Information System*. U.S. Geological Survey. 1930-2009. . Retrieved 2010-09-12.

[4] U.S. Geological Survey. National Hydrography Dataset high-resolution flowline data. The National Map (http://viewer.nationalmap.gov/viewer/), accessed March 16, 2011

- Busch Gardens - San Luis Rey River (http://www.buschgardens.org/swc/wetlands/sd_county_wetlands/san_luis_rey.htm)
- Project Clean Water - San Luis Rey River (http://www.projectcleanwater.org/html/ws_san_luis_rey.html)

Article Sources and Contributors

Reservoir *Source*: http://en.wikipedia.org/w/index.php?title=Reservoir *Contributors*: Aatox, Acalamari, Alansohn, Allstarecho, AmericanLeMans, Amniarix, Andonic, Arjayay, Arthena, Astirmays, Awickert, B.d.mills, BL Lacertae, Babakathy, Bassbonerocks, Billinrio, Blehfu, Bobo192, Brougham96, Camw, Can't sleep, clown will eat me, CanadianLinuxUser, Carnildo, Cassowary, Cctoide, Ceranthor, Chl, Chris j wood, ChrisCork, Ciphers, CommonsDelinker, Courcelles, Csmart287, CustardJack, Cyfal, DFS454, DVD R W, Davandron, Deror avi, DidiWeidmann, Dj Capricorn, Dnj710, Docu, Donarreiskoffer, Drunken Pirate, ENeville, ESkog, Em3rald, Emerson7, Eranjenes2, Erik9, Ewen, Ezhiki, FayssalF, Fieldday-sunday, Fifth Fish Finger, Floul1, Frehley, Fribbler, Frokor, Funandtrvl, Fæ, GLaDOS, Gail, GavinSharp, Gcm, GeoWriter, Geoff Bonanno, Geonarva, Gersenes, Ghirlandajo, Gilliam, HarlandQPitt, Hertz1888, HexaChord, Hike395, Hires an editor, Hmains, Ian n-s, Infrogmation, Ingolfson, J.Gifford, JBellis, JCDenton2052, JForget, Jacksieber, JamesAM, Jeffrey Mall, JetLover, Jim.henderson, Jonpollnow, Jose Icaza, Jpbowen, KVDP, Kaiba, Karlos87, Kbthompson, Kimse, Kjetil r, Kjkolb, Kostisl, Krj373, Kschwerdt514, Levineps, Lobob5, Logan, Lysy, Mailseth, Makeemlighter, Malathos, Marcia Wright, Martial75, Mattgirling, Mattisse, MaxSem, MeekMark, Metropolitan90, MichaelMaggs, Minghong, Mircea cs, Moanzhu, Moreau1, Muijz, Muéro, Narson, Newell Post, Nigholith, Nihiltres, Nixsun01, Nk, NortyNort, Nshimbi, Ohdear15, Oismiffy, OlEnglish, Old Moonraker, Orphan Wiki, Osarius, Pengo, Pete ess, Peter gk, Peterlewis, Philip Trueman, Pigsonthewing, Planetjanet, Pointillist, PopUpPirate, Prabinepali, Prashanthns, Qfl247, RJFJR, Radiojon, Rajasekhar1961, Ravedave, Redmegtheavenger, Rich Farmbrough, Robina Fox, Roxi2, Rruss, S4n1HS22WMX695In, Samnz05, SchuminWeb, Sean the Spook, Severnapark, Sfahey, Shannon1, Sole Soul, Squids and Chips, Svetovid, Swanml, Tabletop, Tedickey, TenOfAllTrades, Tgeairn, TheGerm, Thingg, Thw1309, Tide rolls, Tills, Tom harrison, Totse-vandal, Tranquil demon, Trevyn, Trilby*foxglove, Tsuite, Túrelio, Uncle Dick, Unschool, VPliousnine, Vdegroot, Vegaswikian, Velella, Versus22, Vhorvat, VinceBowdren, Wafulz, Walrusfunk, Wavelength, Whosasking, Wikid77, Wikijmms, William Avery, YixilTesiphon, 250 anonymous edits

Bank (geography) *Source*: http://en.wikipedia.org/w/index.php?title=Bank_%28geography%29 *Contributors*: Altenmann, Awickert, Bertrand Bellet, ErikvanB, Fiftytwo thirty, Look2See1, Michael Hardy, Moreau1, Noles1984, Peter Horn, Sionnach1

Lake Henshaw *Source*: http://en.wikipedia.org/w/index.php?title=Lake_Henshaw *Contributors*: Djrun, Shannon1, Wikidemon

Llwyn-on Reservoir *Source*: http://en.wikipedia.org/w/index.php?title=Llwyn-on_Reservoir *Contributors*: Bardsandwarriors, Docu, Geopersona, Jeremy Bolwell, Lewisdg2000, M-le-mot-dit, MRSC, Malcolma, MarylandArtLover, Mhockey, Thediscreetguy, Trident13, Velella, 1 anonymous edits

North County, San Diego *Source*: http://en.wikipedia.org/w/index.php?title=North_County%2C_San_Diego *Contributors*: 08OceanBeach SD, AThing, BlankVerse, Branddobbe, BronzeAddy, Cooljuno411, Dave Andrew, Dohn joe, Evil saltine, Fratrep, Gill Giller Gillerger, House1090, J04n, Jllm06, Joelwest, Kylephoto760, Look2See1, MalcolmHeardSB, NE2, Neilc, Rick Block, Rnickel, Rschen7754, SchutteGod, Ski09, SoCal L.A., Techieman, WhisperToMe, XR9, Ylee, Zzyzx11, 20 anonymous edits

Palomar Mountain *Source*: http://en.wikipedia.org/w/index.php?title=Palomar_Mountain *Contributors*: Beeblebrox, Branddobbe, Chris the speller, Clutch1, Davemcarlson, David Eppstein, Droll, Durandall05, Edcolins, Gary Jones GJ, Glazed, GrahamHardy, Hmains, Hydrogen Iodide, Itinerant1, Jan.Kamenicek, Jllm06, John of Reading, KGyST, LessHeard vanU, Look2See1, Lotje, Mike Dillon, NE2, Ost316, PhilipKirbyJames, Plbman, R Lee E, RedWolf, Richard Weil, Tylerfinvold, 28 anonymous edits

San Diego *Source*: http://en.wikipedia.org/w/index.php?title=San_Diego *Contributors*: -Midorihana-, 08OceanBeach SD, 11tas, 123wizard123, 16@r, 1958publius, 20bucket, 360view, 4musicipod, 72Dino, A Train, A1%, AMAPO, ATC, Aahndee, AaronRoe, Abennobashi, Ablask77, AbsolutDan, Academic Challenger, Acalamari, Achangeisasgoodasa, Achimowicz, Acntx, Adraeus, Advancedants, AeonicOmega, Aflfinals, Afrancis, Againme, Agamemnon117, AgnosticPreachersKid, Ahacopian, Ahoerstemeier, Airplaneman, Al Silonov, Alansohn, Ale jrb, Alex LaPointe, AlexiusHoratius, Alihoruz, Allstarecho, Alphaboi867, Amazonien, Ambyent, AmeliaElizabeth, Amerique, Amire80, Anb racsh, Anclation, Andre Engels, Andrewia, Andrewlp1991, Andrewphelps, Andrewpmk, Andrwsc, Andy Marchbanks, Andyjsmith, Andyuts, Anetode, Angel caboodle, Anger22, Angr, Anietor, AnnaFrance, AnnaP, Anon134, Antandrus, AntonioMartin, Aoi, Apothecia, Arawk, Arch dude, Arden, Ariusturk, ArnoLagrange, Arpingstone, Aruton, Asdfasdf1231234, Asinensky, Askusinternational, Aspects, Attilios, AuburnPilot, Aude, Aungzayar, Avicennasis, Avril.caballo, AySz88, Aykroyd, Azumanga1, BC Lafferty, Baile01, Bakudai, Balajiprasenna, Bantman, Baptisst, Bassbonerocks, Bbichnev, Bbygrl love, Bcorr, Bdj00, Beeblebrox5000, Beetstra, Ben Ben, Ben Lunsford, BenSanders, Bender235, Benhealy, Bentley4, Berwil 92083, Betterusername, Bewareofdog, Bhavs26, Bhw752k, Big Bird, Bignd500, Bilbrauer, Bill1968, Billwhittaker, Binksternet, Bishop^, Bizzwriter, Black N Red, BlackMeph, Blairmariew, BlankVerse, Blawton777, Blixtra.org, BlkMtn, Blue520, Blueboy96, Boatman, Bobblewik, Bobby H. Heffley, Bobet, Bobo192, Bogsat, Boing! said Zebedee, Bombcar, BorgQueen, Born2cycle, Born2x, Boudiccat, Bovlb, Boznia, Brainyiscool, Branddobbe, Breadandcheese, BrendelSignature, Brgoldsmith, Brian1078, Brianga, Brion VIBBER, Brodey, Brthomas, Bryan A, Bucephalus, Buchanan-Hermit, Buhtelka, Bunnyhop11, Burroughsks88, BuyAMountain, C to the O-R-N, C777, CGorman, CIreland, CJLL Wright, CJLucke, CKBrown1000, CWY2190, Cacetudo, Calaschysm, Calimo, Cambam2009, CambridgeBayWeather, Can't sleep, clown will eat me, CanDo, CanadianLinuxUser, Cantiorix, Capricorn42, Captain-tucker, Carcarroxursox, Carl.bunderson, Carlossuarez46, Carterthedog, Casper2k3, Catalan, Cazort, Cburnett, CdaMVvWgS, Cecole, Celsius1414, Centrx, ChargersFan, CharlesBronson18, Charlesdrakew, Chart123, Cheesesteaks3, Chemical Halo, Chicbicyclist, Choster, Chowbok, Chr.K., Chris 73, Chris Barna, Chris ecorreo@yahoo.com, Chris the speller, Chrisch, Christopher Mann McKay, Chutney379, Cinik, Ckatz, Cleared as filed, Clq, Cmcnicoll, Cntras, Colipon, Collins8857, Commanderaminius, CommonsDelinker, Conversion script, Cool Stuff Is Cool, CoolKid1993, Coolcaesar, Cooljuno411, Corpx, Coulraphobic123, Courcelles, CovenantD, CrazyC83, Crazysunshine, Creamy4, Creashin, Cripwalk2diswun, Crzrussian, Csloomis, Cyanidethistles, Cyrusc, D6, DJ Clayworth, Dachannien, Dale Arnett, Daltnpapi4u, Dananderson, Daringbibliofile, Darklilac, Darth Panda, Darz Mol, Dasani, Davidphogan74, Dbez04, Dbiel, Dcgomez, Dean1970, DeanKeaton, Debigboy, Debresser, Deflective, Deisom, Delkevin, Delldot, Delta G, DeltaQuad, Deltabeignet, Demiurge, DerHexer, Derek.cashman, Derek85, DevOhm, Deydreamer79, Dheerajpetla, Dillard421, Dingar, Diogenes00, Diomidis Spinellis, Discospinster, Dmurray88, Doc Tropics, Doc glasgow, DocWatson42, Docu, Doczilla, Dohn joe, DonDeigo, Donald Albury, Doniago, Donnymo, Doogie18, DopefishJustin, Dorftrottel, Doron, Doseiai2, Doug Coldwell, Dough4872, Dpbsmith, Drasek Riven, Dravecky, Drdr1989, Dschwen, DuncanHill, Durova, Dysepsion, Dzhuo, ERK, Eastlaw, Eco84, Ed g2s, Edchi, Eddstah, Edgecliff89, EditorASC, Edobrichoo, EdwardOConnor, Efrem7, Efyoo2, Eggie5, El C, Elliot hicks, Elpaso2wasilla, Emigrant85, Enlightened explorer, Enviroboy, Enzedbrit, Epbr123, Epolk, Erechtheus, Erianna, Eric Bekins, Esanchezyn, Esperant, Essjay, EugeneZelenko, EurekaLott, Everyking, EvilSing, Excirial, Extra999, FF2010, FUgators33, Facts707, Falcon8765, Faradayplank, Farine, FayssalF, Fertejol, FieldMarine, Figureground, Firsfron, Fixer88, Flauto Dolce, Flibirigit, Flip619, Flowanda, Forehowe, Foreverprovence, FrancoGG, Fransurber2010, Freakdog, Freakofnurture, FredR, Frehley, FrenchIsAwesome, Ftc08, Funnyhat, Fusillijerry86, Fxhomie, G-Man, G. Capo, GDonato, GGreeneVa, Gadfium, Gaius Cornelius, Gateman1997, Gboeing, Gcanyon, GeoGreg, GeoffreyHale, Geographer, Georg 101, GeorgeLouis, Gerritg, Geschichte, Gilliam, Gimmetrow, Glacier109, Glane23, Glen1995, Glenlarson, Godgundam10, Gogo Dodo, Golbez, Gouryella, Graham87, Grandmasterka, GrantBarrett, Greenguy1090, Greenshed, Greg Tyler, Gregorknight, Griffinofwales, Ground Zero, Gudeldar, Gundamforce, Gurch, Guroadrunner, Gwernol, H3h, HADRIANVS, HJ32, Ha ha ha 63, HaeB, Haham hanuka, Hajor, Harryhanlon, Hawaiian717, Hawaiimethodman, Hdt83, Headbomb, Healkids, HeartofaDog, Heatwave, Hemanshu, Hgnsd, Hike395, Hintha, HkCaGu, Hlmcmahon, Hmains, Hottentot, House1090, Howcheng, Howefortunate, Hu12, Hubertfarnsworth, Hubschrauber729, Huntington, HurricaneHugo, Hut 8.5, IAMTHEEGGMAN, IRelayer, IWhisky, IceCreamAntisocial, Ief, Ilovelukeburns, Iman sadeghi, Imdbbbd, InTeGeR13, Infrogmation, Inter-man, Intothewoods29, Invertzoo, Iridescent, IrisKawling, Irnisdcausa, Irontomflint, Ithemachine, Itinerant1, Ivirivi00, J Di, J.delanoy, J3ff, JB82, JForget, JHMM13, JNW, JPolitteCA, JaGa, Jack Cox, JackBNimble, Jackfork, Jacoplane, Jacquese, Jakewaage, JamesAM, JamesB3, Jamiemaloneyscoreg, Jaranda, Jareha, Jason L. Gohlke, Jasongolod, Jaybeatle, Jayron32, Jbruin152, Jcheckler, Jclemens, Jcmenal, Jcortes66, Jeffinmo, Jeffrey O. Gustafson, Jenlavin, Jepler, JeremyA, Jessxenos, Jevansen, Jf3mo, Jh51681, Jhendin, Jhubach1, Jiang, Jim.henderson, Jim1138, Jimbenhar, Jj137, Jj98, Jketola, Jm51, Jmundo, Jnorton7558, Jo7hs2, JoJan, JoSePh, JoanneB, Jobroni21, Johan1298, John K, John of Reading, John254, Johnny Jane, Johntex, Joinarnold, Jojhutton, Jon186, JonnyBoy, Joshuakeria, Joyous!, Jscarreiro, Jsmithcloser, Jsnod, JulianDave, Jusdafax, JustAGal, Justin Eiler, K0001213, KFuller, Kafziel, Kaiserb, Kaisershatner, Kakofonous, Kanaye, Kapheee, Karol Langner, Kaskew, Katieh5584, KatyWatkins, Kbdank71, Kcdc123, Keelm, Keithyork, KellBelle75, Kellen`, Kempachi, Kenyon, KevinJ, Kevinatilusa, Killing sparrows, Kilroy Jenkins, King 4rthur, King Reece rules, Kingjeff, Kingpin13, Kingturtle, Kitch, KnowledgeOfSelf, Koavf, Konman72, Koppenlady, Kramden4700, KrazyCaley, Ktr101, Kubigula, Kumioko, Kuru, Kurykh, Kusma, Kwamikagami, LAX, LAboi84, LOL, Lan56, Laszlo2, LaszloWalrus, Laurenprovesbeauwrong, Law, Lazer Shmuel, Lazulilasher, LeaderMgmt, Leaftye, LegendarySlammer5893, Legoktm, LeoNomis, Leslie Mateus, Levineps, Lfh, Liface, Lightmouse, Lilac Soul, Lilpinoy 82, Lincolnite, Lion666, Little Mountain 5, LittleOldMe, Ljmajer, Localmotion528, Lofty, LongDistance06, Loodog, Look2See1, Looper5920, Lord Bodak, Lord Pistachio, Lorraine LeBeau, Lorrainx, Lotje, Lousyd, Lrd1rocha, Ltruett, Luckycharms, Luna Santin, Luong, Luvbach1, MBisanz, MJCdetroit, MK8, Macabe, Mackerm, Mackin90, Magioladitis, Magnus Manske, Magzour, Mais oui!, Mani1, Manutdglory, Mapster10, Marauder40, Marcpschaefer, Marek69, MarkGallagher, Marksburgess, Marthaaa, Martin Jambon, Martinp23, Marysunshine, Mathpianist93, MattSal, Mattscards, Maverick Leonhart, Maximz2005, MayaSimFan, Mbecker, Mboverload, McDange, McSly, Mcvang, Mdewman6, Medmura118, Megapen, Mehudson1, MelanieN, Melchoir, Mentifisto, Michael Hardy, Michaelsbll, Midgrid, Mike Dillon, Mikeinsd, Mikevegas40, MilFlyboy, Mild Bill Hiccup, Mind21 98, Miquonranger03, Mixtli, Mlaffs, Mmdolbow, ModelFish, Mohrflies, MojaveNC, Molerat, Moncrief, Monterey Bay, Monz, Moreati, Moreau36, Mr. Lefty, MrBojanglesNY, MrHudson, MrPMonday, Mrsmith93309, Mtrisk, MuZemike, Muboshgu, Mudoven, Muhand, Murcielago, Mushroom, MusicInTheHouse, Mxn, Myasuda, N328KF, N96, NE2, NHRHS2010, Nahallac Silverwinds, Nakon, Namit80, Narfers02, Nascar1996, Nataly8, Nathan Johnson, NatureA16, Navy1775, Nehrams2020, Netsnipe, Neutrality, NewEnglandYankee, Newnoise, Nflanny, Nichalp, Nickbillings, Nicksoda21, Nigel7615, NigelR, Nights Not End, Nikai, Nironn, Nishkid64, Nn123645, No Annuities, NoSeptember, Noneskull, Northamerica1000, Northwest, Norvy, NorwegianBlue, Nscaglio, Nterage, NuclearWarfare, Nucleusboy, Nv8200p, OMHalck, Oakshade, Obsidian Soul, OcciMoron, Ohconfucius, Ohnoitsjamie, Oiskas, Okiefromokla (old), Old Guard, OllieFury, Optigan13, Orayzio, Orthographer, OverlordQ, Oxymoron83, P41, PQMark, Pablo-flores, PainMan, Pal046, Palfrey, ParisianBlade, Patricknoddy, Patstuart, Paul A, Paulwade, Pb30, Pbroks13, Pearle, Pentawing, Pepper, Pepperdine24, Pepsicrazedkid, Peptuck, PerryPlanet, Perspective, Peruvianllama, Peter135, PeterJohnson, Pfranson, Pgk, Phil Bordelon, Phil Boswell, Philip Baird Shearer, Philip Trueman, Philscherrer1980, Phspiller, PinchasC, Piperdown, Pisceandreams, Plasma east, PochWiki, Pol098, Polaron, Ppayne, Prefect, Prodego, Profoss, Prom3th3an, Proofreader77, Pseybold, Psychris, Puddingman11493, Pylambert, Quadell, Quantpole, Quickymartnot, Qwell the pell, R'n'B, RHCPJedi, RJN, RadiantRay, Radiokid1010, RainbowOfLight, Ram-Man, Ramirez72, RandomP, Randomgbear, RandorXeus, Raw weaver, Rayasmith, RazorICE, Razz tip, Rebelduder69, Red Hair Bow, RedRollerskate, Redvers, Reg619, Rettetast, RexNL, Rhatsa26X, Rich Farmbrough, Richard Myers, Richardthemagical, Rick Block, RickK, RickyCourtney, Rim Pirate, Rjwilmsi, Rjwoer, Rklawton, Rlhuffine, Rmhermen, Rnickel, Roadnottaken, Rockero, Rodhullandemu, Rogerd, Ronaldmolina20, Ronbo76, Rorschach, Rossmwilson, Rowlan, RoyBoy, Rpeh, Rrburke, RufusTeleStrat, Runefrost, RunningFromSatan, Runtime, Rustynuts, Rw11, RxS, Ryankindelan, Ryanpm4545, Rzwilling, SDX, SNIyer12, SPUI, Sadalmelik, Sade, Sagaciousuk, Sallyrob, Salondemaria, Sam Korn, Samaritan, Samba6566, Samhuddy, San diego fan, San2frango, SanDiego619CA, Sanchom, Sandieagogogo, Sandiegodude79, Saxifrage, Scarian, Schcambo, Schmiteye, SchnitzelMannGreek, SchuminWeb, Schzmo, Sciprogrammer, Scoopaloop, ScottDavis, Scotthatton, Scottsdalecom, Scriberius, Sdsoc, Sdzman73, Sean WI, SeanMack, Seattlenow, SecondSlip, Sedna10387, Seja430, Seth Ilys, Seven Days, Sfmusicfan1, Shadowlynk, Shanes, Shasta6, ShelfSkewed, Shell Kinney, Shereth, Shirulashem, Shoeofdeath, Short Verses, Shortride, Shyguy1991, Sillyfolkboy, Silverhelm, SimonD, SimonP, Sintonak.X, Sjakkalle, SkarmCA, Skew-t, Ski09, Skidmore-13, Skionxb, Skk646, Skomorokh, Slapshot579, Slaxson, Slffea, Slowking Man, Smarterchild7, Smashville, Smaug123, Smm650, SmthManly, Smultiplication, Snowyknit, SoCal L.A., Socal gal at heart, Socalres, Soltras, SoundGod3, SpK, Space185, SpaceFlight89, Spellcast, Spike Wilbury, Spikestiffnippleswifebeater, Spitfire, Spongie555, Squiderick, Srich32977, Stan Shebs, Stangoldsmith, Staxringold, Stayclassy, Steel, Stemonitis, Stenun, Stephenb, Stephensuleeman, Stevertigo, Stewartadcock, Streltzer, Stroppolo, Stubblyhead, Student7, Subtropical-man, Sumsum2010, Superflush, Surferchik46, SusanLesch, Syghost, Syvanen, Szulc12, Szyslak, T.J.V., TFOWR, THEN WHO WAS PHONE?, TPK, TRBP, TVfanatic2K, Tameemms, Tangerine Cossack, Tassedethe, Tbhotch, Techman224, Teej, Teflon johnny, Template namespace initialisation script, That Guy, From That Show!, Thatguyflint, The Big Down, The Epopt, The Infamous Q, The Phoenix Enforcer, The Rambling Man, The Red, The Thing That Should Not Be, The Universe Is Cool, The josh17, The359, The762x51, TheCoffee, TheNewPhobia, TheOtherJesse, TheRingess, Thebigboi, Theconroy, Thekillerclover, Theresa knott, Thewellman, Thomas Larsen, Thomasmallen, Thompsontough, Thortful, ThurstonWetsack, Tibetan Prayer, Tiddly Tom, Tide rolls, Tijuanagringo, TimBentley, TimShell, Timbratcher, Timc, Timo Honkasalo, Titoxd, Tmlm1ke, Tmurdoch, Tokakekeke, Toki, Tom Danson, TomasBat, TommyBoy, TonyTheTiger, Toomuchhiking, Tostie14, Touchdown Turnaround, Toymao, Tpbradbury, Track Legs, Traf-quake, Travelforever, Tresiden, Trevor MacInnis, Trewells, Treygiles, Trident13, TriniMuñoz, Trojan51, Trusilver, Tryingtomakethingsright, Ttony21, Tuspm, TutterMouse, Twas Now, Tyler, Udibi, Ufwuct, Ulaireminya, Ulric1313, Un sogno modesto, Uris, User2004, Utelprob, Utezduyar, Vanished 6551232, Vanished User 1004, Vclaw, Vegaswikian, Vegaswikian1, Veinor, Ventusa, Venusantrius, Veraladeramanera, Verbivorous, Vercillo, Versus22, Vorear, Vote4luke, Vrable, Vrenator, Vrysxy, W guice, WHS, WHeimbigner, Walrusonion, Walter Humala, WarthogDemon, Watsurnameandurnumber, Wavehunter, Wavelength, Wayne Schroeder, Wd40gdw, Wdfarmer, Weatherman78, Weeliljimmy, Welsh, WereSpielChequers, Weregerbil, Wesheath35, Wesley M. Curtus, Western Pines, Westmt01, WhisperToMe, Whorchatasoto, Wiithlove, Wiki alf, Wiki.Tango.Foxtrot, Wikicali00, Wikieizor, Wikipediarules2221, Wikipedpad, Will Beback, WillC, William Allen Simpson, Williamv1138, Wizzard2k, Wj32, Woohookitty, Worldenc, Wwarhurst, Wwoods, X-factor, X10guy, XJamRastafire, XLerate, Xeno, Xinoph, Xsportscoalitionsd, Xyxyboy, Yath, Yellowboy06, Yellowdesk, Ygoloxelfer, Ylee, Youndbuckerz, Yuckfoo, Yuje, Ywong137, ZachPruckowski, Zachlipton, Zeitgeist248, Ziggurat, Zink Dawg, Zoe, Zscout370, Zzuuzz, Zzyzx11, מ בר., 2322 anonymous edits

Groundwater *Source*: http://en.wikipedia.org/w/index.php?title=Groundwater *Contributors*: A. B., Aharlan, Aitias, Alansohn, Ali K, Angela, Anlace, Antandrus, Anthonares, Anthony Appleyard, Aoi, Arakunem, ArcticFrog, Argyriou, Arthena, Badgernet, Barek, Basar, Bayerischermann, Beetstra, Beland, Bigbluefish, Bobo The Ninja, Bobo192, Bryan-Oddfellow, CDM2, CSWarren, Calltech, Caltas, Calvin 1998, CarolGray, Ccgrimm, Charles Dickens, Chase.alton3, Chicco3, Chriswaterguy, Chroja, Chuunen Baka, Civil Engineer III, Clayton orgles, Clifftreyens, Cmdrjameson, Cp111, Cwiweb, DJKostaki, DVD R W, Daniel Collins, Dariusofthedark, David Levy, Dawn Bard, DeansFA, Discospinster, Djdutch, Doniago, Drm310, Dzubint, Eastlaw, Elfino, Ellisthion, Emc2, Entropy, Epbr123, Epipelagic, ErikDunsing, Everyking, EvilPizza, Fabiform, FayssalF, Fieldday-sunday, Fiveless, Forstbirdo, Fourdee, Frankenpuppy, Frazzydee, FreplySpang, Fresno2000, Gcm, Gdphdb, GentlemanGhost, Geographer, Geologyguy, Glen, Glenn, Gob Lofa, Gralo, GregE625, Gunnernett, Hdt83, Hollowtools, Huzzahmaster018, Indiawaterportal, Infrangible, J.delanoy, J04n, JForget, JTN, JamesBWatson, Jarbon, Jauerback, Jauhienij, Jeff G., Jessa marie, JodyB, John254, Johnathonsmithe, Jon Nevill, Jrdioko, Julesd, Jusdafax, Jóna Þórunn, Kateshortforbob, KatieDOM, Katoa, Kbh3rd, Khalid hassani, Koiplaats, Krellis, KrisK, Ksyrie, Kuhlman, Kungming2, Kwells1989, L Kensington, Leafyplant, Lilac Soul, Loltamere, Look2See1, Mac, Madhero88, Mahjongg, Mailseth, Malo, Marisa269x, Marshman, Martin451, Materialscientist, Maurreen, Megiddo1013, Mentifisto, Mikenorton, Mintleaf, Mirv, Mnemo, Moeron, Monfornot, Monsoon Waves, Moreau1, Mygerardromance, Nakon, Natalie Erin, NawlinWiki, NcRomance, NellieBly, Nick, Nick123, Omicronpersei8, Omnicog, Optimist on the run, Otisjimmy1, Oxbox, Oxymoron83, Pacula, Panyd, Parutakupiu, Pearle, Perchedaquifer, Persian Poet Gal, Phantomsteve, Pietrow, Pill0031, Pinethicket, Pinky sl, Pollinator, Pupster21, Qfl247, Quintote, Qxz, R.J.Oosterbaan, R.steenhard, R123triplehhhfan123, Raven4x4x, RedWolf, Redvers, Reinyday, Rjwilmsi, RocketRoger, Salgueiro, Salmar, Samw, Sean.hoyland, Seaphoto, Shiftchange, Shizane, Shlcrest05, Shrumster, Siim, Skater, Slawojarek, Smack, Smallverm, Some jerk on the Internet, Sonett72, Soumyadeep2208, SpaceFlight89, SparrowsWing, Steven Weston, Suntag, Synthiss, TARATHILIEN, Tellyaddict, The Anome, The High Fin Sperm Whale, The Thing That Should Not Be, Theo Pardilla, Tide rolls, Tom harrison, Tschel, USchick, Ubergeekguy, V.narsikar, Vervin, Vincentcayabyab, Viridae, Voice of reason993, Vsmith, WadeSimMiser, WarthogDemon, Watertruth, Wavelength, Weetoddid, WikiLaurent, Wikipelli, Wilson44691, Wimt, Wrp103, Xiahou, Xionbox, Zeizmic, Zuejay, 585 anonymous edits

San Luis Rey River *Source*: http://en.wikipedia.org/w/index.php?title=San_Luis_Rey_River *Contributors*: Asiaticus, Chris the speller, ClamDip, Hmains, Jllm06, Ken Gallager, Kingturtle, Kjkolb, Perdelsky, Shannon1, Stepheng3, YourEyesOnly, 5 anonymous edits

Image Sources, Licenses and Contributors

Image:Taiwan JungHua Dam.JPG *Source*: http://en.wikipedia.org/w/index.php?title=File:Taiwan_JungHua_Dam.JPG *License*: unknown *Contributors*: User:Vegafish

Image:Lakevyrnwysummer.jpg *Source*: http://en.wikipedia.org/w/index.php?title=File:Lakevyrnwysummer.jpg *License*: unknown *Contributors*: User:Sean the Spook. Original uploader was Sean the Spook at en.wikipedia

Image:Stocks Reservoir.jpg *Source*: http://en.wikipedia.org/w/index.php?title=File:Stocks_Reservoir.jpg *License*: unknown *Contributors*: User:Karlos87

Image:GibsonR.jpg *Source*: http://en.wikipedia.org/w/index.php?title=File:GibsonR.jpg *License*: unknown *Contributors*: User:Citypeek

Image:Hydroelectric dam.svg *Source*: http://en.wikipedia.org/w/index.php?title=File:Hydroelectric_dam.svg *License*: unknown *Contributors*: User:Tomia

Image:KupferbachStauseeAachen.jpg *Source*: http://en.wikipedia.org/w/index.php?title=File:KupferbachStauseeAachen.jpg *License*: unknown *Contributors*: User:Túrelio

Image:Llyn Brianne spillway.jpg *Source*: http://en.wikipedia.org/w/index.php?title=File:Llyn_Brianne_spillway.jpg *License*: unknown *Contributors*: User:Velela

Image:volta lake.jpg *Source*: http://en.wikipedia.org/w/index.php?title=File:Volta_lake.jpg *License*: unknown *Contributors*: NASA

Image:Lake Kariba.jpg *Source*: http://en.wikipedia.org/w/index.php?title=File:Lake_Kariba.jpg *License*: unknown *Contributors*: Martin H., Mehmet Karatay, Paobac, TheDJ, TommyBee, 1 anonymous edits

File:Kuekenhoff Canal 002.jpg *Source*: http://en.wikipedia.org/w/index.php?title=File:Kuekenhoff_Canal_002.jpg *License*: unknown *Contributors*: User:Deepak

File:Namoi-River-sand-bank.jpg *Source*: http://en.wikipedia.org/w/index.php?title=File:Namoi-River-sand-bank.jpg *License*: unknown *Contributors*: User:Floybix

file:Llwyn-on Reservoir - geograph.org.uk - 148765.jpg *Source*: http://en.wikipedia.org/w/index.php?title=File:Llwyn-on_Reservoir_-_geograph.org.uk_-_148765.jpg *License*: unknown *Contributors*: Docu

File:Oceansidepier.jpg *Source*: http://en.wikipedia.org/w/index.php?title=File:Oceansidepier.jpg *License*: unknown *Contributors*: User:Alessandriana

File:North County San Diego.png *Source*: http://en.wikipedia.org/w/index.php?title=File:North_County_San_Diego.png *License*: unknown *Contributors*: User:Cooljuno411

File:Banded juvenile California Least Tern.jpg *Source*: http://en.wikipedia.org/w/index.php?title=File:Banded_juvenile_California_Least_Tern.jpg *License*: unknown *Contributors*: Linda Tanner

Image:Del-Mar-Race-Track.jpg *Source*: http://en.wikipedia.org/w/index.php?title=File:Del-Mar-Race-Track.jpg *License*: unknown *Contributors*: Helenaak, Intersofia, Mrrxx, NapoliRoma, Stan Shebs

Image:Oceanside Beach Panorama.jpg *Source*: http://en.wikipedia.org/w/index.php?title=File:Oceanside_Beach_Panorama.jpg *License*: unknown *Contributors*: Randy OCH

Image:Grand01.jpg *Source*: http://en.wikipedia.org/w/index.php?title=File:Grand01.jpg *License*: unknown *Contributors*: User:Dcmcgov

Image:DMTC July 2008.jpg *Source*: http://en.wikipedia.org/w/index.php?title=File:DMTC_july_2008.jpg *License*: unknown *Contributors*: User:Achen33

Image:Lakesanmarcos.jpg *Source*: http://en.wikipedia.org/w/index.php?title=File:Lakesanmarcos.jpg *License*: unknown *Contributors*: User:Taylorj661

Image:Oceanside Pier.jpg *Source*: http://en.wikipedia.org/w/index.php?title=File:Oceanside_Pier.jpg *License*: unknown *Contributors*: Bill Newman

Image:TwinPeaksPoway1.jpg *Source*: http://en.wikipedia.org/w/index.php?title=File:TwinPeaksPoway1.jpg *License*: unknown *Contributors*: Perdelsky

Image:Palomar Observatory.jpg *Source*: http://en.wikipedia.org/w/index.php?title=File:Palomar_Observatory.jpg *License*: unknown *Contributors*: Quadell, SoWhy, Tylerfinvold

File:SD Montage.jpg *Source*: http://en.wikipedia.org/w/index.php?title=File:SD_Montage.jpg *License*: unknown *Contributors*: User:Nehrams2020, User:SusanLesch

File:Flag of San Diego, California.svg *Source*: http://en.wikipedia.org/w/index.php?title=File:Flag_of_San_Diego,_California.svg *License*: unknown *Contributors*: Original uploader was Zscout370 at en.wikipedia

File:Seal Of San Diego, California.svg *Source*: http://en.wikipedia.org/w/index.php?title=File:Seal_Of_San_Diego,_California.svg *License*: unknown *Contributors*: Original uploader was Zscout370 at en.wikipedia

File:San_Diego_County_California_Incorporated_and_Unincorporated_areas_San_Diego_Highlighted.svg *Source*: http://en.wikipedia.org/w/index.php?title=File:San_Diego_County_California_Incorporated_and_Unincorporated_areas_San_Diego_Highlighted.svg *License*: unknown *Contributors*: Arkyan

file:Usa edcp location map.svg *Source*: http://en.wikipedia.org/w/index.php?title=File:Usa_edcp_location_map.svg *License*: unknown *Contributors*: User:Uwe Dedering

File:Red pog.svg *Source*: http://en.wikipedia.org/w/index.php?title=File:Red_pog.svg *License*: unknown *Contributors*: Anomie

File:Kumeyaay.jpg *Source*: http://en.wikipedia.org/w/index.php?title=File:Kumeyaay.jpg *License*: unknown *Contributors*: Kosigrim, Nehrams2020

File:San Diego Murillo.jpg *Source*: http://en.wikipedia.org/w/index.php?title=File:San_Diego_Murillo.jpg *License*: unknown *Contributors*: Bartolomé Esteban Perez Murillo

File:San-diego-mission-chuch.JPG *Source*: http://en.wikipedia.org/w/index.php?title=File:San-diego-mission-chuch.JPG *License*: unknown *Contributors*: User:Dmadeo

File:Alonzo Horton.jpg *Source*: http://en.wikipedia.org/w/index.php?title=File:Alonzo_Horton.jpg *License*: unknown *Contributors*: Dananderson, Infrogmation, 1 anonymous edits

File:Guide Book of the Panama California Exposition.jpg *Source*: http://en.wikipedia.org/w/index.php?title=File:Guide_Book_of_the_Panama_California_Exposition.jpg *License*: unknown *Contributors*: Dananderson, Durova, Nehrams2020, 1 anonymous edits

File:San Diego-Tijuana JPLLandsat.jpg *Source*: http://en.wikipedia.org/w/index.php?title=File:San_Diego-Tijuana_JPLLandsat.jpg *License*: unknown *Contributors*: NASA/JPL/NIMA

File:NormalHeightsSmall.jpg *Source*: http://en.wikipedia.org/w/index.php?title=File:NormalHeightsSmall.jpg *License*: unknown *Contributors*: Tone (talk) Original uploader was Monotone at en.wikipedia

File:SanDiego panorama.jpg *Source*: http://en.wikipedia.org/w/index.php?title=File:SanDiego_panorama.jpg *License*: unknown *Contributors*: U.S. Navy photo by Mass Communication Specialist Seaman Christopher K. Hwang

File:Magnify-clip.png *Source*: http://en.wikipedia.org/w/index.php?title=File:Magnify-clip.png *License*: unknown *Contributors*: User:Erasoft24

File:blacks surfer.jpg *Source*: http://en.wikipedia.org/w/index.php?title=File:Blacks_surfer.jpg *License*: unknown *Contributors*: Original uploader was Agc is me at en.wikipedia

File:PacificBeach2.jpg *Source*: http://en.wikipedia.org/w/index.php?title=File:PacificBeach2.jpg *License*: unknown *Contributors*: Tim Shell. Original uploader was TimShell at en.wikipedia

File:Torrey Pines State Park Valley.jpg *Source*: http://en.wikipedia.org/w/index.php?title=File:Torrey_Pines_State_Park_Valley.jpg *License*: unknown *Contributors*: User:Nauticashades

File:San Diego skyline against smoke from wildfires Oct 2007.jpg *Source*: http://en.wikipedia.org/w/index.php?title=File:San_Diego_skyline_against_smoke_from_wildfires_Oct_2007.jpg *License*: unknown *Contributors*: Kat Miner from San Diego, USA

File:US Navy 091008-N-9761H-041 Vice Adm. D.C. Curtis and Intelligence Specialist 3rd Class Ryan Paigo cut a cake during a ceremony in observance of Hispanic American Heritage Month.jpg *Source*: http://en.wikipedia.org/w/index.php?title=File:US_Navy_091008-N-9761H-041_Vice_Adm._D.C._Curtis_and_Intelligence_Specialist_3rd_Class_Ryan_Paigo_cut_a_cake_during_a_ceremony_in_observance_of_His *License*: unknown *Contributors*: Docu, Sanandros

File:P1000734-crop.jpg *Source*: http://en.wikipedia.org/w/index.php?title=File:P1000734-crop.jpg *License*: unknown *Contributors*: User:SusanLesch

File:FA18CHornetOverSanDiegoNov08.jpg *Source*: http://en.wikipedia.org/w/index.php?title=File:FA18CHornetOverSanDiegoNov08.jpg *License*: unknown *Contributors*: U.S. Navy

File:Qualcomm headquarters.jpg *Source*: http://en.wikipedia.org/w/index.php?title=File:Qualcomm_headquarters.jpg *License*: unknown *Contributors*: Coolcaesar at en.wikipedia

File:LaJollaSkyline.jpg *Source*: http://en.wikipedia.org/w/index.php?title=File:LaJollaSkyline.jpg *License*: unknown *Contributors*: 08OceanBeach SD, Courcelles, FlickreviewR, Lymantria, 1 anonymous edits

File:United States Department of the Navy Seal.svg *Source*: http://en.wikipedia.org/w/index.php?title=File:United_States_Department_of_the_Navy_Seal.svg *License*: unknown *Contributors*: US Army Institute Of Heraldry

File:Sdsumain.jpg *Source*: http://en.wikipedia.org/w/index.php?title=File:Sdsumain.jpg *License*: unknown *Contributors*: Original uploader was Geographer at en.wikipedia

File:Geisel library.jpg *Source*: http://en.wikipedia.org/w/index.php?title=File:Geisel_library.jpg *License*: unknown *Contributors*: Original uploader was Four at en.wikipedia

File:MuseumofManSD08.jpg *Source*: http://en.wikipedia.org/w/index.php?title=File:MuseumofManSD08.jpg *License*: unknown *Contributors*: User:Nehrams2020

File:QualcommChargersRams.JPG *Source*: http://en.wikipedia.org/w/index.php?title=File:QualcommChargersRams.JPG *License*: unknown *Contributors*: Original uploader was Nehrams2020 at en.wikipedia

File:Petco Park altitude.jpg *Source*: http://en.wikipedia.org/w/index.php?title=File:Petco_Park_altitude.jpg *License*: unknown *Contributors*: Bryan, FlickreviewR, Uris

File:Westfield Horton Plaza.jpg *Source*: http://en.wikipedia.org/w/index.php?title=File:Westfield_Horton_Plaza.jpg *License*: unknown *Contributors*: User:Vrysxy

File:San Diego City Council chambers.jpg *Source*: http://en.wikipedia.org/w/index.php?title=File:San_Diego_City_Council_chambers.jpg *License*: unknown *Contributors*: Bengt Nyman

File:JerrySandersByPhilKonstantin.jpg *Source*: http://en.wikipedia.org/w/index.php?title=File:JerrySandersByPhilKonstantin.jpg *License*: unknown *Contributors*: User:Philkon

File:I-5 South in San Diego.jpg *Source*: http://en.wikipedia.org/w/index.php?title=File:I-5_South_in_San_Diego.jpg *License*: unknown *Contributors*: Original uploader was Jepler at en.wikipedia

File:OverCoronadoSanDiegoAug07.jpg *Source*: http://en.wikipedia.org/w/index.php?title=File:OverCoronadoSanDiegoAug07.jpg *License*: unknown *Contributors*: Doug Letterman at http://www.flickr.com/photos/dougletterman/

File:Flag of Spain.svg *Source*: http://en.wikipedia.org/w/index.php?title=File:Flag_of_Spain.svg *License*: unknown *Contributors*: Anomie

File:Flag of Brazil.svg *Source*: http://en.wikipedia.org/w/index.php?title=File:Flag_of_Brazil.svg *License*: unknown *Contributors*: Anomie

File:Flag of the Philippines.svg *Source*: http://en.wikipedia.org/w/index.php?title=File:Flag_of_the_Philippines.svg *License*: unknown *Contributors*: Aira Cutamora

File:Flag of the United Kingdom.svg *Source*: http://en.wikipedia.org/w/index.php?title=File:Flag_of_the_United_Kingdom.svg *License*: unknown *Contributors*: Anomie, Mifter

File:Flag of Afghanistan.svg *Source*: http://en.wikipedia.org/w/index.php?title=File:Flag_of_Afghanistan.svg *License*: unknown *Contributors*: 5ko, Ahmad2099, Antonsusi, Avala, Bastique, Dancingwombatsrule, Dbenbenn, Denelson83, Domhnall, Duduziq, F l a n k e r, Fry1989, Gast32, George Animal, Happenstance, Herbythyme, Homo lupus, Klemen Kocjancic, Koefbac, Kookaburra, Lokal Profil, Ludger1961, MPF, Mattes, Myself488, Neq00, Nersy, Nightstallion, Orange Tuesday, Rainforest tropicana, Reisio, Rocket000, Sojah, Tabasco, Zscout370, 28 anonymous edits

File:Flag of South Korea.svg *Source*: http://en.wikipedia.org/w/index.php?title=File:Flag_of_South_Korea.svg *License*: unknown *Contributors*: Various

File:Flag of Mexico.svg *Source*: http://en.wikipedia.org/w/index.php?title=File:Flag_of_Mexico.svg *License*: unknown *Contributors*: User:AlexCovarrubias

File:Flag of Australia.svg *Source*: http://en.wikipedia.org/w/index.php?title=File:Flag_of_Australia.svg *License*: unknown *Contributors*: Anomie, Mifter

File:Flag of the People's Republic of China.svg *Source*: http://en.wikipedia.org/w/index.php?title=File:Flag_of_the_People's_Republic_of_China.svg *License*: unknown *Contributors*: User:Denelson83, User:SKopp, User:Shizhao, User:Zscout370

File:Flag of the Republic of China.svg *Source*: http://en.wikipedia.org/w/index.php?title=File:Flag_of_the_Republic_of_China.svg *License*: unknown *Contributors*: 555, Bestalex, Bigmorr, Denelson83, Ed veg, Gzdavidwong, Herbythyme, Isletakee, Kakoui, Kallerna, Kibinsky, Mattes, Mizunoryu, Neq00, Nickpo, Nightstallion, Odder, Pymouss, R.O.C, Reisio, Reuvenk, Rkt2312, Rocket000, Runningfridgesrule, Samwingkit, Sasha Krotov, Shizhao, Tabasco, Vzb83, Wrightbus, ZooFari, Zscout370, 72 anonymous edits

File:Flag of Ghana.svg *Source*: http://en.wikipedia.org/w/index.php?title=File:Flag_of_Ghana.svg *License*: unknown *Contributors*: Benchill, Fry1989, Henswick, Homo lupus, Indolences, Jarekt, Klemen Kocjancic, Neq00, OAlexander, SKopp, ThomasPusch, Threecharlie, Torstein, Zscout370, 4 anonymous edits

File:Flag of Russia.svg *Source*: http://en.wikipedia.org/w/index.php?title=File:Flag_of_Russia.svg *License*: unknown *Contributors*: Anomie

File:Flag of Poland.svg *Source*: http://en.wikipedia.org/w/index.php?title=File:Flag_of_Poland.svg *License*: unknown *Contributors*: Anomie, Mifter

File:Flag of Japan.svg *Source*: http://en.wikipedia.org/w/index.php?title=File:Flag_of_Japan.svg *License*: unknown *Contributors*: Anomie

File:Shipot.jpg *Source*: http://en.wikipedia.org/w/index.php?title=File:Shipot.jpg *License*: unknown *Contributors*: User:USchick

Image:AlapahaRiver2002.jpg *Source*: http://en.wikipedia.org/w/index.php?title=File:AlapahaRiver2002.jpg *License*: unknown *Contributors*: Stewart Tomlinson, Florida

File:High plains fresh groundwater usage 2000.svg *Source*: http://en.wikipedia.org/w/index.php?title=File:High_plains_fresh_groundwater_usage_2000.svg *License*: unknown *Contributors*: User:Kbh3rd

File:Groundwater flow.png *Source*: http://en.wikipedia.org/w/index.php?title=File:Groundwater_flow.png *License*: unknown *Contributors*: Original uploader was KrisK at en.wikipedia

Image:MiddleSpring.JPG *Source*: http://en.wikipedia.org/w/index.php?title=File:MiddleSpring.JPG *License*: unknown *Contributors*: User:Citypeek

Image:Limestone building with pollution.jpg *Source*: http://en.wikipedia.org/w/index.php?title=File:Limestone_building_with_pollution.jpg *License*: unknown *Contributors*: Ali K, J.-H. Janßen, 1 anonymous edits

Image: Sanluisreyriverphoto1.jpg *Source*: http://en.wikipedia.org/w/index.php?title=File:Sanluisreyriverphoto1.jpg *License*: unknown *Contributors*: Perdelsky

Image:Sanluisreyriverphoto2.jpg *Source*: http://en.wikipedia.org/w/index.php?title=File:Sanluisreyriverphoto2.jpg *License*: unknown *Contributors*: Perdelsky

Image:Sanluisreyriverphoto3.jpg *Source*: http://en.wikipedia.org/w/index.php?title=File:Sanluisreyriverphoto3.jpg *License*: unknown *Contributors*: Perdelsky

Image:Lakehenshaw1.jpg *Source*: http://en.wikipedia.org/w/index.php?title=File:Lakehenshaw1.jpg *License*: unknown *Contributors*: Perdelsky

Image:Lakehenshaw2.jpg *Source*: http://en.wikipedia.org/w/index.php?title=File:Lakehenshaw2.jpg *License*: unknown *Contributors*: Perdelsky

Image:Lakehenshaw3.jpg *Source*: http://en.wikipedia.org/w/index.php?title=File:Lakehenshaw3.jpg *License*: unknown *Contributors*: Perdelsky

image:San_Luis_Rey_River_High_Water_2010.jpg *Source*: http://en.wikipedia.org/w/index.php?title=File:San_Luis_Rey_River_High_Water_2010.jpg *License*: unknown *Contributors*: User:EboMike

Printed by Books on Demand GmbH, Norderstedt / Germany